Extraterrestrial Life

Antonino Del Popolo

Extraterrestrial Life

We are not alone

 Springer

Antonino Del Popolo
Department of Physics and Astronomy
University of Catania
Catania, Italy

ISBN 978-3-031-83496-7   ISBN 978-3-031-83497-4   (eBook)
https://doi.org/10.1007/978-3-031-83497-4

This book is a translation of the original Italian edition "La vita extraterrestre" by Antonino Del Popolo, published by Springer Nature Switzerland AG in 2025. The translation was done with the help of an artificial intelligence machine translation tool. A subsequent human revision was done primarily in terms of content, so that the book will read stylistically differently from a conventional translation. Springer Nature works continuously to further the development of tools for the production of books and on the related technologies to support the authors.

Translation from the Italian language edition: "La vita extraterrestre" by Antonino Del Popolo, © The Author(s), under exclusive license to Springer Nature Switzerland AG 2025. Published by Springer Nature Switzerland. All Rights Reserved.

This Springer imprint is published by the registered company Springer Nature Switzerland AG
The registered company address is: Gewerbestrasse 11, 6330 Cham, Switzerland

If disposing of this product, please recycle the paper.

Introduction

There are two possibilities. We are alone in the Universe or we are not. Both are shocking
  Arthur C. Clarke

Many years ago, in a hill town closer to Africa than to the rest of Italy, on a clear and warm summer evening, I was looking at the sky, as I often did. Since then, the stars have been much better visible than they are today. Light pollution is making it increasingly difficult to observe the wonders that stand out in the celestial vault. On that beautiful evening, a clear, milky strip was visible, a bridge suspended in the firmament: the *Milky Way*, which owes its name to one of the stories of Greek mythology. I already knew that the milky stripe is nothing other than the manifestation of the stars of our galaxy located near the galactic disk. I imagined walking on that sort of suspended bridge and being able to move between distant worlds, passing with one step from one star system to another. Physical reality limits us; it does not allow us to reach the speed of light, and it does not allow us to reach those distant worlds unless we have very advanced technologies and a very long time available. Imagination makes us travel everywhere in our galaxy or the Universe, in infinitesimal times. My imagination galloped and led me to see other worlds in that endless white vastness, other star systems with their planets, and their inhabitants busy with their daily activities. Perhaps on some of those planets it was night and another living being, lost in that immensity, looked toward the sky and imagined, like me, that somewhere someone was observing the firmament and asking himself the same questions. Those very youthful times (few years had passed since Apollo 11 had brought the first men to the moon in 1969) were fervent times of ideas related to the cosmos and interplanetary, interstellar or intergalactic travel. In those years, TV films such as *Space 1999* and *UFO,* created by Gerry and Sylvia Anderson, were widely followed by kids. In *Space 1999,* the Moon detached from its orbit and wandered into space, and the approximately three hundred inhabitants of *Alpha Base,* traveling through the cosmos, encountered new planets and new forms of life, some hostile, some friendly. At the time, it was thought that man was destined for interstellar travel in the near future. In addition to science fiction books and films, space science has developed owing to a whole series of missions. In 1957, the Soviet Union launched Sputnik 1, followed by flights with human crews such as the one on which the dog Laika was. A few years later, in 1961, the cosmonaut Yurij Gagarin (Fig. 1) was the first human to fly in the outer space.

Apollo 11, as remembered, brought the first men to the moon in 1969. Various other missions were carried out, with the goal of reaching, flying over, or photographing the planets of our solar system.

The general mood was that man would soon conquer space and, most interestingly, that he might come into contact with extraterrestrial civilizations. It seems that the human species is led to think that there is life everywhere, and this has been the case since ancient times. Aristarchus of Samos, and Epicurus, developing ideas of Leucippus and Democritus, in the fourth and third centuries BC hypothesized that the universe was full of worlds suitable for hosting life. Lucretius, in *De rerum natura* in the first century BC,

Fig. 1   Jurij Gagarin. (Credit: Finnish Museum of Photography)

wrote that it was absurd to think that an infinite space was created only in our world. This thesis was attacked by Plato, Timaeus, and Aristotle, who described the uniqueness of the Earth. In contrast, Giordano Bruno, who was burned alive in 1600, imagined that there was an infinity of other worlds, other planets, populated by conscious beings. In 1752, Voltaire wrote the philosophical novel *Micromegas,* in which he described extraterrestrial life forms. The main character, Micromega, 120,000 feet tall, lives on a planet in the star Sirius from which he begins journeys to other worlds. From an astronomical point of view, the structures observed on the Moon were interpreted as the result of the work of an extraterrestrial civilization; similarly, there was speculation about the existence of life on Mars. These speculations were fueled by the observations of the astronomer Schiaparelli and by several of his articles. The natural structures he observed were wrongly translated into English: instead of *chanels,* the term *canals were used,* implying the ideas they were built by living beings. The American Percival Lowell supported the idea that these canals were artificial canals. Owing to Lowell's successful books written between 1895 and 1908 and the famous science fiction novel *The War of the Worlds* by H.G. Wells published in 1898, the term *Martian* became synonymous with extraterrestrial, and it was convinced that there was conscious life on Mars. William Sinton's observations in the late 1950s prompted him to write articles in which he talked about the existence of plants on Mars. Only with observations with more accurate instruments was it understood that Lowell's ideas were incorrect, and finally, the images of Mariner 5 in 1965 and Mariner 9 in 1971 definitively closed all speculation on *Martians.* The debate was particularly hot after the publication of the book *Of the Plurality of Worlds: An Essay* by William Whewell in 1853. For a decade, the existence of other planets was discussed, and the fact that they were not observed led to the conclusion that the Earth was the only planet in the Universe and that life existed only on it. In 1950, Enrico Fermi, during a lunch with some colleagues, reasoning about the enormous number of stars in the Universe and the fact that there were no signs of the existence of extraterrestrial civilizations, posed the famous question "where is everyone?", a question encapsulating what is now known as the *Fermi paradox.*

This paradox arrives at the conclusion that we are alone. Thus, over the centuries, people oscillated between conflicting theses. The ubiquity of life on Earth naturally leads us to think that the birth of life is automatic, mixing the concepts of the genesis of life with those of its expansion. Before Louis Pasteur, people believed in *spontaneous generations.* Dirty rags can breed mice, and rotting meat can create maggots. Therefore, it was thought that generating life from the nonliving was easy. Pasteur's experiments revealed that these were erroneous beliefs. The *Miller–Urey experiment* also contributed to showing how it is not easy to create life from nonliving material. The enthusiasm of the 1960s and the following decade, in which people imagined that man would soon travel and conquer new worlds in our Universe and would encounter extraterrestrial life, has faded slightly, both for the reasons mentioned and for the understanding that the distances in Universe are enormous. The closest star, Proxima Centauri, is 4.2 light years away. With current technology, with which, for example, the *Parker Solar Probe* can travel, it would take approximately 7100 years to travel there. *Yuri Milner,* an Israeli IT magnate, philanthropist and physicist, Stephen Hawking, cosmologist and astrophysicist, and Mark Zuckerberg founded the *Breakthrough Initiatives* in 2016, generating a research and engineering project that has a series of ambitious goals. The program is divided into several projects. *Breakthrough Listen* will include an effort to search for artificial radio or laser signals in more than 1,000,000 stars. A side project called the *Breakthrough Message* is an effort to create a message representative of humanity and planet Earth. The *Breakthrough Watch project* aims to identify and characterize Earth-sized rocky planets around Alpha Centauri and other stars within 20 light years from Earth. Also included in the project is sending a mission to Saturn's moon, Enceladus. The Breakthrough Starshot project aims to reach Proxima Centauri in the past few decades via space sails, which are a few square meters long and a few µm thick and weigh a few grams, pushed by a powerful laser emitted from Earth. With this technique, we would aim to reach Alpha Centauri in 20 years. Thinking back to when I was a child, I realize how many steps forward have been made, but as we know, knowledge is like a bubble; the more it expands, the more it comes into contact with new surfaces: the more we know, the more we have left to know. The fantasies that led me to think that the Universe was full of living creatures more or less strange and different from us, borrowed from the science fiction of my youth, lost their vividness compared with our current scientific knowledge. To date, we have no proof of the existence of other life in the entire Universe apart from that on Earth, despite the many efforts made to have news, through some radio signal, that we are not alone in the Universe. Hopes were revived by the discovery of *extrasolar planets*, namely, *exoplanets*. Compared with the 1960s and 1970s, today, we know that the Universe is full of planets. In our galaxy alone, there could be a hundred billion of them, and current estimates suggest that there is at least one rocky, terrestrial type, for every five stars such as ours, i.e., something like six billion. These numbers grow enormously if hundreds of billions of galaxies are considered in the universe. Life, primitive or otherwise, could exist on one of these planets. The hope that extraterrestrial life may exist has even given rise to a new science, *astrobiology* (or *exobiology*), a term introduced in 1955 by Otto Struve. This science has grown substantially since the 1960s, and many centers worldwide address life in general and extraterrestrial life in particular. Astrobiology is about answering questions such as the following: What is life? A seemingly trivial question but one which is not easy to answer. How did life originate on Earth? Does extraterrestrial life exist? How can we find it, and in the case of evolved life, how can we communicate with extraterrestrial civilizations? Our nature has led us for millennia to think that we are special beings, that the Universe had been built for us, a vision with man at the center, the so-called *anthropocentric vision*, which history has taught us to be wrong, and that our role in the Universe is not central. We discovered, owing to Copernicus, that the Earth is not at the center of the solar system and that the Sun is not at the center of the galaxy and that our galaxy is not at the center of the Universe. Centuries had to pass before this truth, the *Copernican principle*, became part of our culture. This principle also pushes us to think that since we are not special beings, it is probable that other forms of life, or other civilizations, exist in the Universe. This extrapolation, although natural, may not be true. We do not know how difficult it is for life to take root on a planet, and even if the number of planets is very large, we cannot be sure that life is present in some other remote corner of the Universe. We only know about terrestrial life, and extrapolating, inferring that other life forms exist in the Universe, could lead us down the wrong path. So all we have to do is follow the dictates of science and continue to scan space with the technological means we have in search of some signal sent by some other civilization, and at the same time use our technology to study our solar system using space probes, and as for the planets of other stars, we could study their atmospheres which could contain signs of the action of some form of life, *biosignatures*. Could we answer our question about the existence of extraterrestrial life in this way? As we will see below, this is likely the case. If we were lucky, we might find some life in our solar system, for example, in some satellites of the giant planets or on Mars. Otherwise, we would need to study the planets of the stars in our galaxy or, if possible, those of all galaxies in the Universe. If we consider the number of stars, it is enormous: ten thousand billion, probably five or ten times more than all the grains of sand on all the Earth's beaches. Obviously, there is no way to study them all, but even considering only the stars in our galaxy, a few hundred billion, they are too many to study. We must limit ourselves to the nearest stars, and in any case, they are an enormous number. It is precisely this enormity that gives us good hope that there is life on some of the many exoplanets.

Competing Interests The author has no competing interests to declare that are relevant to the content of this manuscript.

Contents

1

Children of the Stars

We are made of star-stuff
Carl Sagan

In March 1806, an unusual object had fallen from the sky above the village of Valence in southern France. It was collected by farmers, who took it to scientists for analysis. They discovered that it contained water, organic matter, and matter made up of carbon and other elements. Several years later, the meteorite was brought to the laboratory of the famous scientist Louis Pasteur, who studied it to understand whether it contained any form of life. He found no life but confirmed the presence of organic material. Meteorites such as Valence constitute 5% of all meteorites and are called *carbonaceous chondrites* because they contain spheroidal regions called chondrules and carbon and carbon compounds. These meteorites formed at the origin of the solar system and are important in the search for extraterrestrial life that is supposed to be made up of organic matter. In 1969, a similar meteorite, the *Murchinson meteorite*, fell in Australia. The analysis revealed basic elements on which life is based: *amino acids*. A spontaneous question is what chemical elements and phenomena are found in space. The most accredited theory for the formation of the Universe is the *Big Bang theory*, which states that the Universe formed approximately 13.8 billion years ago from a very small region that began to expand rapidly. Approximately 3 min after the Big Bang, light elements, such as hydrogen, helium and some other light elements, formed. The heavier elements formed much later in stars.

© The Author(s), under exclusive license to Springer Nature Switzerland AG 2025
A. Del Popolo, *Extraterrestrial Life*, https://doi.org/10.1007/978-3-031-83497-4_1

However, helium not only formed in the primordial Universe but also continuously formed in stars, and it was precisely in one of them, the Sun, that Joseph Norman Lockyer discovered it in 1868. Its name, helium, comes from the name of the Sun in Greek, *helios*. In stars beyond helium, carbon and all the other heavy elements are formed. The formation of carbon in stars (and then that of other elements) was discovered by Fred Hoyle. In stars, three helium nuclei try to unite to form carbon (in the so-called *three-alpha process*), but the probability that 3 helium atoms combine to form carbon-12, in its known form, is almost zero. Hoyle assumed that in the named process, in which three helium atoms form a carbon atom, it is necessary for the carbon nucleus to have a very particular energy. The large amount of carbon in the universe, which results in the existence of carbon-based life forms, was proof that this particular energetic state of carbon, now called the *Hoyle state*, had to exist. On the basis of this idea, Hoyle predicted the values of the energy and other parameters of the compound state in the carbon nucleus formed by three alpha particles (i.e., helium nuclei). This state was shown in subsequent experiments. The great importance of this discovery lies in the fact that owing to the presence of this energy level, it is possible to explain the large production of carbon-12 present in red giants (a stage of stellar evolution described in Chap. 7), which acts as a ring of conjunction between the synthesis of the lighter and heavier chemical elements. Without this energy level, the production of carbon-12 would be approximately 10^7 times lower, as the three-alpha process could not proceed via resonance.

Regarding the existence of this very particular state of carbon, Hoyle said:

> *A commonsense interpretation of the facts suggests that a super intellect has monkeyed with physics, chemistry and biology and that there are no blind forces worth speaking about in nature. The numbers one calculates from the facts seem to me so overwhelming as to put this conclusion almost beyond question*

and, in his 1983 book *The Intelligent Universe*, Hoyle wrote:

> *The list of anthropic properties, apparent accidents of a nonbiological nature without which carbon-based and hence human life could not exist, is large and impressive*

In other words, Hoyle preceded the idea of *fine-tuning*, namely, that in the Universe, some parameters are finely regulated, that is, if, in the Universe, the constants of nature did not have the value that they have, life as we know it would not exist. A point of view that was later taken up by Robert H. Dicke, according to whom some forces, such as gravity and electromagnetism, must be finely regulated for life to exist in the Universe. These points of view led

Brandon Carter to the formulation of the *anthropic principle* in 1973, but to discuss these aspects, another book would be needed.

Carbon is the main element involved in the development of life. It is capable of building long chains through chemical reactions, binding mainly to hydrogen, which is fundamental for life. It is the realm of organic chemistry, which has this name because many organic compounds are produced by living beings. For example, a diamond is made up of many carbon atoms bonded together. Carbon can form carbon dioxide, meaning that a carbon atom can share two electrons with oxygen atoms that form carbon dioxide. Carbon also forms methane, which consists of one carbon atom and four hydrogen atoms around it. The early Earth's atmosphere consisted mainly of these two gases. As we previously observed, carbonaceous chondrites that fall from the sky contain carbon; therefore, the space also contains carbon. Stars can exist precisely owing to the transformation, in nuclear processes, of hydrogen into helium, helium into carbon and so on up to iron. This element cannot be fused to form heavier elements because it requires energy instead of producing it. Once a star has an iron core, it is performed. The inner part collapses, and the star explodes, resulting in a *supernova*, and the outer parts are hurled into space, forming a *supernova remnant* (Fig. 1.1).

Fig. 1.1 Supernova remnant. (Credit: NASA, ESA, and STScI)

All other elements that are heavier than iron are formed during the explosion phase. These supernovae disseminate the elements we know throughout space. The first stars that were born in the Universe contained only light elements (hydrogen and helium) and then formed heavier elements within them and scattered them throughout space. From this material, other stars and other supernovae were formed. It therefore took billions of years before the Universe was equipped with enough heavy materials to also form the disks from which, as we will see, planets are born. This explains why the Universe is 13.8 billion years old and why our solar system is only 4.5 billion. Life also requires a whole series of elements before it can appear. Then, carbonaceous chondrites and complex organic molecules formed in the interstellar clouds. The large presence of carbon in space is also demonstrated by the behavior of the star R Coronae Borealis. The star changes brightness within a period of a few weeks. This is due to the emission of large clouds of soot, a mixture of organic and inorganic substances. Oxygen, which is also formed in stars, is the source of the energy used by life. It is very present on Earth. It accounts for 21% of the air by volume and is the most common element in the Earth's mantle. Oxygen is heavier than carbon but can form fewer compounds than it does. It is very reactive, and this is fundamental to life, at least life after 2 billion years. Before, life did not use oxygen, and the Precursor LUCA (Last Universal Common Ancestor) probably did not exchange oxygen or gas with the immediate environment. When oxygen binds with other elements, energy is released, which is capable of supporting life in an organism. Oxygen with hydrogen forms water, another fundamental element for life. Oxygen and carbon form carbon dioxide. When a compound containing carbon comes into contact with oxygen in the form of energy (e.g., heat), molecular oxygen splits into atomic oxygen, which, when combined with carbon, gives rise to carbon dioxide. The opposite process, i.e., the liberation of oxygen from carbon dioxide, requires the activity of living organisms. It is thought that in exoplanets, which we discuss later, there cannot be molecular oxygen or ozone unless living beings exist. Therefore, the search for extraterrestrial life on exoplanets goes hand in hand with the search for oxygen in their atmosphere. For example, when plants appear on Earth, owing to the chlorophyll photosynthesis practiced by plants, they are able to split carbon monoxide into oxygen and carbon via solar energy, filling the Earth's atmosphere with oxygen, which allows the development of animal life. A 2024 study revealed that, on Jupiter's satellite Europa, approximately one thousand tons of oxygen are generated every 24 h.

This, together with other observations that we discuss in Chap. 5, suggests that there may be life on Europe. There is also oxygen on Ganymede, but it is not thought to be of biotic origin; rather, it is produced as a result of the incident radiation on the surface, which determines the splitting of water ice molecules present on the surface of the satellite into hydrogen and oxygen. Proteins, fundamental components of life, contain not only carbon, oxygen and other elements but also 16% nitrogen. Therefore, nitrogen is also essential for terrestrial life. Nitrogen binds to three oxygen atoms, forming ammonia, which is a gas present wherever there is decomposition of organic matter. Furthermore, 78% of the air is composed of molecular nitrogen. Another important element for life is sulfur. It is one of the essential elements for vegetables and is a component of several amino acids and vitamins. One of the hypotheses of the origin of life, as we will see in Chap. 4, is based on sulfur and iron in the depths of the oceans. In 2017, researchers from the University of Trento reported that the iron and sulfur groups underlying the enzymes necessary for life may have floated above primordial seas approximately 4 billion years ago. They are produced by primitive molecules activated by ultraviolet light. Therefore, the ingredients for life could be the Sun together with the iron–sulfur groups. Another element of interest is silicon, which, as we will see in Chap. 9, has been proposed as a substitute for carbon; i.e., the existence of life on the basis of this element has been proposed. Silicon has intermediate characteristics between those of metals and nonmetals. A silicon compound was found in the Arizona meteor crater, which shows its presence in space. However, by bonding to hydrogen, silicon cannot form long chains or ring structures. Silicon is much more versatile. Metals also have fundamental importance for life. For example, iron, in addition to being one of the initiating elements of life on Earth, is one of the constituents of hemoglobin, which is used to transport oxygen in the body. Crabs use copper for the same function. Sodium and potassium are essential for the functioning of the nervous system, and calcium forms teeth and bones. Zinc and other metals are also essential for proper functioning. The atoms that make up living beings are relatively few compared with all those found in the periodic table and make up our universe: carbon, hydrogen, nitrogen, oxygen, phosphorus, and sulphur, with the addition of others such as sodium, calcium, potassium and fluorine. We also obtained traces of iron, iodine, magnesium, zinc, selenium, copper, manganese, chromium, and molybdenum. All these elements were formed in stars, whose explosions scattered them around the cosmos. It is just as Carl Sagan claimed: *we are made of star stuff.*

2

Towards Complexity

Even if we had a billion monkeys who knew how to type, the possibility that they
would be able to write correctly, during a period equal to the age of the universe,
even a single Dante's triplet is almost nil...
Francis Crick

2.1 What Is Life?

Since I was a child, I have always been fascinated by living things, but I believe
all human beings are. I truly liked cats, and I asked myself many questions
about their behavior, how they are made, and the difference between a cat and
an inanimate object. The answer then seemed banal: the cat responded when
I caressed it; it responded to external stimuli, whereas an inanimate object did
not do all this. This question of mine about what life is and which for me had
a banal solution is a question that men have been asking themselves for thousands of years, and even today, we do not have a precise definition. For example, for Aristotle, living beings, unlike inanimate beings, had three types of
soul: vegetative, animal, and rational (in the case of human beings). George
Erns Stahl and others defined the doctrine of *vitalism* in the seventeenth century. According to vitalists, living organisms differ from nonliving organisms
because they are equipped with nonphysical elements, and they also believe
that the organic matter, of which we are made, cannot be derived from the
inorganic matter, of which nonliving beings are made.

In reality, today, we know that these ideas were wrong and that, for example, inorganic material can be transformed into organic material. Attempts to
define what life is certainly did not stop with Stahl. Many scientists have

A. Del Popolo, *Extraterrestrial Life*, https://doi.org/10.1007/978-3-031-83497-4_2

attempted to define life, but if we read a modern biology book instead of reading a short definition, we find a list of properties of life: *order, growth, reactions to stimuli, reproduction, evolution, metabolism, autopoiesis, etc.* This list, like others, fails to capture all the characteristics of living beings. To provide some examples, we can consider crystals that are organized and grow simultaneously, but we do not believe that they are living beings. If we consider the bacteria, they may be inactive for long periods, but they are not dead. The response to stimuli is limited not only to living organisms but also to some human-designed machines. Reproduction does not define a living being either. Some living beings cannot reproduce on their own. For example, mules are sterile but are living beings. The *Turritopsis nutricula* or *immortal jellyfish* does not reproduce, but we consider it alive. The other aspect, evolution, in the sense of storing information in molecules such as DNA or RNA, which we will talk about and transmitting information to offspring, is a unique ability of living beings; for this reason, many biologists have tried to define life on the basis of the concept of evolution. For Carl Sagan, the element that characterizes life is evolution, and his statement according to which life is a *system capable of evolving through natural selection* is well known. This definition also includes artificial life, i.e., computer entities that replicate and evolve under virtual selective pressures. Currently, most people do not accept the idea that this is real life. Beyond this problem, there is another practical problem in recognizing the activation of an evolutionary process because observing the system over a period of thousands of years is necessary. Furthermore, sterile individuals such as mules, celibates, and, in general, a single individual cannot reproduce; therefore, they do not enter the game of evolution and therefore would be nonliving beings. In contrast, according to Sagan's definition, viruses are living beings, but much of the scientific community does not consider them as such. Defining life seems difficult; not even evolution, although intimately linked to life, is sufficient for its recognition. Gerald Joyce of the Scripps Research Institute and consultant to NASA's exobiology program gave a working definition that generalized Sagan's: a living being is *a self-sustaining chemical system capable of experiencing Darwinian evolution*. This definition is also not without drawbacks. For example, a parasitic worm that lives in a person's intestine, despite having all the genetic information to reproduce, cannot do so without the cells of the intestine from which it takes the energy necessary for its survival. Therefore, he would not be a living being. What can we say relatively to *metabolism*, i.e., the set of chemical transformations that are dedicated to vital maintenance within the cells of living organisms, relating to the distinction of living beings from nonliving ones? For Margaret Boden, an expert in artificial intelligence, metabolism distinguishes natural living beings from those with artificial life and from viruses. However, if metabolism refers to the exchange of matter and energy, then it is also a property of fire or tornadoes. Another concept to which life has been linked is *autopoiesis*, that is, the ability of a living being to maintain its individuality. Humberto Maturana and Francisco Varela identified a living as an *autopoietic system*. The definition of life as an autopoietic system given in 2000 by Varela is a little too abstract, so much so that it considers artificial life as real life and has the problem that it eludes the evolutionary aspects of life.

Defining what life is seems to be a hopeless undertaking. It seems to be a clear concept for everyone, but defining it is complex. This situation is reminiscent of the one in which Saint Augustine found himself in his attempt to define time. In his *Confessions,* he wrote:

If no one asks me, I know what it is; if I wish to explain it to him who asks, I do not know.

The definitions given over the years by various scientists have the characteristic that they can be either too concrete or too abstract. In the first case, false negatives are favored, i.e., a living system can be judged as nonliving. In the second case, the opposite happens, and a nonliving system can be judged as living. Especially for those who are trying to find evidence of life in space, it is essential to define what life is. This is important because it is necessary for the recognition of any type of life, not just the one we know and that exists on Earth. The problem is that apparently, there does not seem to be a simple marker, or rather a *biomarker*, that allows us to recognize life and whose absence leads us to discard its existence. Perhaps the problem is conceptual because it is not easy to define something that we have not yet fully understood. This discussion applies to life in general, but if somewhere in our solar system or on some planet outside it we encountered terrestrial life, we would probably be able to recognize it using some of its unique characteristics, such as DNA. If we came across a form of life different from that of Earth, we might not realize that it is life. While it is theoretically possible to send probes to some satellites of the solar system on which there could be life and repeat analyses such as those of the Viking probes on Mars in 1976, for planets outside our solar system, we could only study their atmospheres and try to draw conclusions. For example, departure from the characteristic equilibrium of life influences the atmospheres of planets. The departure from equilibrium is a necessary condition for the appearance of life, which can then change atmospheres and lead to gas concentration relationships. Another marker of the presence of life is a high level of order. However, the latter is not easy to evaluate. All this happens if life is born on the surface of a planet, but if we deal with underground life, the aforementioned markers may not be observed. Detecting signs of life on a distant planet is obviously complicated. However, we must not forget that until almost 30 years ago, we did not even know if there were extrasolar planets, and although discovering a planet around a star was not easy, we succeeded. This invites us to look to the future of the search for life in space in an optimistic way.

2.2   The Engine of Life

Although, as we have seen, it is difficult to give a general definition of what life is, there are markers that, excluding particular cases such as viruses, allow us to tell us whether we are in the presence of a living object or not. Life is characterized by particular reactions described by Linus Pauling in his book *General Chemistry*. A plant or animal can reproduce and generate offspring belonging to the same species. Chemical reactions can also modify a nonliving object such as a mineral, but it is not capable of reproducing like a living being. Living beings, as we have seen, have a *metabolism*: they take in food, and through chemical reactions, they obtain energy and expel waste products. Pauling realized that there were difficulties in defining life and that there were borderline cases such as viruses. Although plant viruses can replicate via the genetic material of plants, they cannot move or ingest food and are metabolized. They cannot be considered life forms, but in any case, they are complex beings. A characteristic of living beings is that they are made up of cells. A bacterium consists of only one cell. Plants and animals are composed of many cells and different types of cells. A cell is essentially made up of water and *proteins*, which are very large molecules. The molecules have a weight equivalent to thousands to hundreds of thousands of carbon atoms. Our body is made up of different types of proteins, each of which has different vital functions. In turn, proteins are composed of *amino acids* and organic acids. Organic compounds come in two forms: one is a mirror image of the other. We speak of right-handed molecules (and indicate them with D) and left-handed molecules (and indicate them with L). To visualize this aspect, we can think of our hands, the right one and the left one. They are the same, but they do not overlap. The image of the right hand in the mirror is the same as that of the left hand, and vice versa. Amino acids exist in both D and L forms, except for one amino acid, *glycine*. On the other hand, proteins contain only the L form of amino acids, which is strange to say the least. If the chemical reactions that produce the chemical compounds of living organisms occur randomly, one would expect living organisms to be made up of half L-form amino acids and half D-form amino acids, but this is not the case. Proteins are composed of long chains of amino acids, which are called *polypeptide chains.* There are techniques that allow us to count the number of polypeptide chains that form a protein; for example, four polypeptide chains are needed to form *hemoglobin.* The chemical properties of a protein depend on its three-dimensional structure, but chemists, until 1950, did not know how to determine the three-dimensional shape of the chains that make up a protein. Using X-ray diffraction, Linus Pauling was able to determine the structure of a polypeptide chain. They were made up of carbon, nitrogen, hydrogen and oxygen atoms, which formed filaments with a spiral structure. The filaments were wound upward in a clockwise direction, and the others were wound in the opposite direction. However, the amino acids in the chains were all L-type. A problem that remains unsolved is how we went from amino acids and proteins to life. We discuss studies on this topic in a later chapter. In 1951, James D. Watson and Francis Crick discovered the structure of the molecule of interest, DNA, or *deoxyribonucleic acid.* This molecule is the only one capable of replicating itself and, as a consequence, allows an organism to grow and produce offspring. Watson, considering Pauling's discovery of the protein helix, concluded that DNA also has a similar structure. Determining the spatial structure of an organic molecule is fundamental to understanding its behavior. X-ray diffraction images of DNA indicated its structure but not its complete structure. However, Watson's idea, as mentioned, was that the structure was shaped like a helix. This molecule contains a large amount of information about DNA. The problem was to understand how this information was encoded and how it was read to then build a living being. Chemical analysis revealed that the molecule was composed of sugar, phosphorus, and nitrogen. Many studies have confirmed that the structure of DNA is a double helix. Watson and Crick were awarded the Nobel Prize in 1962 for discovering that structure. The two strands of the helix are wrapped around each other, and each strand is made up of repeating units, called *nucleotides*, made up of a sugar (*deoxyribose*), one or more phosphate groups and a nitrogenous base. There are four nitrogenous bases in DNA: *thymine* (T), *cytosine* (C), *adenine* (A), and *guanine* (G). Since there are 4 nitrogenous bases, the number of nucleotides is also 4. The individual's genetic code is written in DNA by a combination of these four molecules. Adenine (A) binds only to thymine (T), and cytosine (C) binds only to guanine (G). Nucleotide sequences form *genes* that contain complete information for a given property (Fig. 2.1).

The complete sequence of nucleotides that make up our genetic heritage is called the *genome,* which is made up of approximately 50,000 genes,

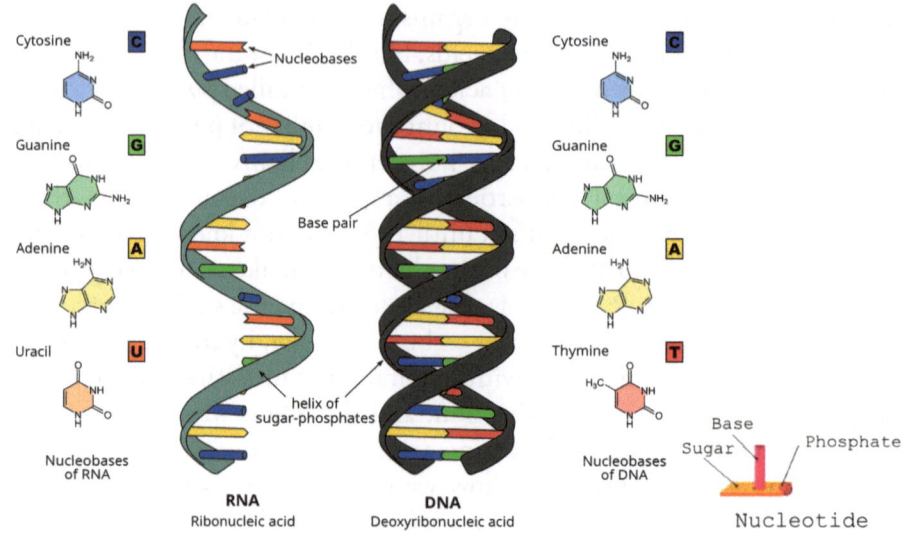

Fig. 2.1 (a) Nucleid acids: RNA and DNA, (b) Nucleotide. (Credits: A. Del Popolo: God or Science?: Is Science Denying God? World Scientific)

corresponding to a length of 3.5 billion "letters". A question that arose at this point was as follows: how do DNA characteristics pass from one generation to the next? To simplify the discussion, the DNA stretches along its entire length and opens like a zipper. The information written on the DNA is copied onto another nucleic acid similar to DNA, called *RNA,* and consists of only one strand. This RNA, which contains the DNA information transcribed onto it, is called *messenger RNA* (mRNA). Messenger RNA removes of the cell nucleus and is read by organelles called *ribosomes, which are* also made up of another RNA called *ribosomal RNA* (rRNA). In this way, each of the DNA strands is synthesized and finally reunited to form a copy of the original DNA. In the process of reading messenger RNA, proteins are produced. The 4 nucleotides of DNA are used to encode the 20 amino acids used to construct proteins. Coding of the 20 amino acids can occur because amino acids are determined by triplets of nucleotides, with 64 possible combinations of triplets. **More details about DNA and protein production can be found in Appendix A.**

DNA is contained in the nucleus of the cell and in the *mitochondria,* organelles that function as energy centers of the cells. DNA "tells" the cell what to do. It tells the cell how to metabolize sugar to extract energy, how to eliminate waste products, and when and how to divide to give rise to daughter cells. DNA contains all the information that determines the characteristics of a species. When two individuals mate, the DNA contained in the genes determines how information about the individual will be passed on to the next generation. DNA is therefore the basic molecule of life, and it is natural to ask the question of what and how it originated on the primordial Earth. From simple random reactions between atoms? The complexity of the origin of life and DNA led Francis Crick to conclude

> *The mechanism necessary to make the genetic code operational, which is universal, is too complex to have arisen in one fallen swoop. An honest man, armed with all the knowledge available to us now, could only state that, in some sense, the origin of life appears at the moment to be almost a miracle, so many are the conditions which would have had to have been satisfied to get it going.*

As we will see in the following chapters, no one has managed to explain the origin of life and DNA on Earth, but considerable progress has been made in understanding it. Moreover, life appeared only a few hundred million years after the Earth had cooled. Crick together with his colleague Orgel thought that this period was not sufficient for the birth of life. Starting from speculations on the genetic code, the two were convinced of the difficulty and improbability of the formation of DNA in that short period. According to them, life arose over a longer period of time somewhere in the Universe and was then transported by an intelligent life form capable of traveling in space. This hypothesis is usually indicated with the term *directed panspermia*. Another, more natural, possibility is that life was transported to Earth by celestial objects. The idea of *panspermia*, supported by some physicists, has lost some of its initial appeal; however, it has not yet been ruled out, and therefore, we will discuss it in the next chapter.

3

Are We Extraterrestrials?

The universe is a pretty big place. If it is just us, appears to be an awful waste of space
Carl Sagan

Svante Arrenhius proposed a new theory of the formation of life in 1906. According to him, life was not born on Earth but had arrived from space. According to him, unicellular organisms would have travelled the sidereal spaces for millions of years to reach the Earth. The push that would make them travel in space would be the radiation pressure of the starlight. Once on a planet, they could give rise to life and higher life forms guided by Darwinian evolution. The idea that led Arrenhius to these conclusions was the observation that terrestrial microorganisms could reach the stratosphere and reach space. It was therefore also possible that from a planet on which life was present, it had reached the Earth or other parts of the Universe. Arrenhius' hypothesis is called *panspermia* (from the Greek, "common seed"). A few decades later, in the 1960s, Carl Sagan was attracted to this idea. To confirm the hypothesis of Arrenhius, microorganisms were found in the stratosphere. Sagan built a mathematical model based on the radiation pressure of sunlight and the gravitational force of the Sun and described, together with Josif S. Shklovskii, the results in a 1966 book: *Intelligent Life in the Universe*. If these two forces are equal, the organism remains in space. If the gravitational force is greater than the radiation force, the organism will fall onto the Sun. If the radiation pressure of light is greater than the gravitational force, the organism will move away from the solar system into the interstellar space. The model also provided the dimensions of these microorganisms, which would have had a radius between 0.2 and 0.6 thousandths of a millimeter, typical dimensions of the spores of fungi and bacteria. These microorganisms would have taken a few years to leave the solar system and a few tens of thousands of years to reach, for example, Proxima Centauri, the closest star. If a microorganism with a radius of less than two thousandths of a millimeter was close enough to the solar system, it would enter it, and in its motion, it could deposit itself on planets, such as the Earth. Sagan also calculated the distance from which the microorganisms that could have given rise to life on Earth left. They would have originated from some planetary systems at a distance of no more than six thousand light years. The next question was whether a microorganism could survive thousands of years in space and then, upon arriving on a planet, become active and spread throughout life. In the vicinity of the Sun, ultraviolet and X-ray radiation can destroy unprotected microorganisms; if they had found themselves inside a meteor, perhaps they would have been able to resist. Interstellar spores can resist cosmic rays for hundreds of millions of years. Another problem is that the probability of a microorganism falling on a planet whose dimensions are very small compared with the scale of the movement of the microorganism is very low. Large numbers of microorganisms are ejected into our galaxy over long periods of time.

3.1   Bjurakan

In 1971, a conference was organized in Bjurakan, Armenia, which was intended to address the topic of communication with extraterrestrial intelligent beings. Carl Sagan, Frank Drake, whom we will talk about later, and Phillip Morrison, who had written the article *Searching for Interstellar Communications,* attended the conference. Crick was also there, interested in life in space. After the studies on DNA, he arrived at the conclusions described in the previous chapter about the extreme difficulty that DNA was formed by random processes on Earth. Additionally, in his opinion, the Earth was too young for such a rare event as the birth of DNA to have occurred. Furthermore, it was necessary to add that life had formed only a few hundred million years after Earth had formed. From scientific dating techniques, it was concluded that life was already present on Earth 3.9 billion years ago, i.e., 600 million years after the formation of the Earth. In agreement with Hoyle, Crick thought that this was a decidedly short time for DNA to form. Therefore, to him, panspermia was a way out of the problem. The DNA could have formed somewhere else where it would have had time.

Nowadays the argument that 600 million years are a too short period for the origin of life is contested, since we do not know if in the presence of catalysts, the random formation of RNA and DNA is fast or not. Crick together with Leslie Orgel presented the idea of the extraterrestrial origin of life, whereas Sagan played the role of devil's advocate, defending the position that this was not possible, owing to the radiation problem, and rejected Crick's idea that the formation of DNA was a very rare event. The conference did not reach any clear conclusions for or against panspermia. Crick and Orgel, starting from speculations on the genetic code, were also convinced, as already mentioned, of the difficulty and improbability of the formation of DNA. The production of a living system, according to them, was a very rare event, but once it started somewhere in the Universe, it could have been spread by an intelligent life form capable of traveling in space. This hypothesis is usually indicated with the term *directed panspermia*.

3.2   ALH84001

Is there evidence that life on Earth could have arisen owing to panspermia? To answer this question, let us remember an event that occurred in 1984: the discovery of a meteorite at Allan Hills in Antarctica, today known as ALH84001 (i.e., Allan Hills 84001) (Fig. 3.1).

The researcher who found the meteorite initially thought it was an aerolite. Brought to the United States, the meteorite was ignored for a decade, until after NASA's 1976 Viking missions to Mars. The analyses revealed that the air

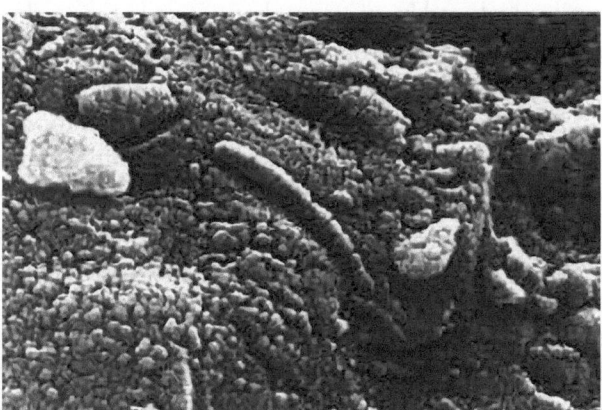

Fig. 3.1   chain structure morphologically similar to biological organisms on a fragment of the ALH 84001 meteorite, seen under an electron microscope. (Credit: NASA)

bubbles inside it had the same composition as the atmosphere of Mars, and in 1993, it was concluded that the meteorite had originated on Mars. The hypothesis about its arrival on Earth is that, 15 million years ago, Mars was hit by an asteroid and that fragments of Martian soil were thrown into space. ALH84001 was one of those fragments that, after wandering for 15 million years, fell to Earth approximately 13,000 years ago. The gas in ALH84001 was in perfect agreement with the Viking data. In 1994, the meteorite was delivered to David McKay of NASA's Johnson Space Center. These studies led to the conclusion that the rock was formed 4.5 billion years ago under the surface of Mars. Organic molecules are *polycyclic aromatic hydrocarbons* that form when a microorganism dies. Electron microscopy revealed that ALH84001 contains magnetite and pyrrhotite, which are also produced by microorganisms. In addition, canals and small tubes that may have been created by living organisms were observed. Taken together, McKay's group concluded that the meteorite contained Martian life, and in 1996, the president of the USA Bill Clinton officially announced the discovery on television and spoke of "a potentially epochal discovery". The evidence provided by Viking that there was water on Mars approximately 3.6 billion years ago, and therefore, it was possible that there was life, gave greater credibility to the conclusions of McKay's group. The 1996 announcement reinvigorated the idea of panspermia. What happened to the rock on Mars, its subsequent drift for 15 million years in space, and its fall to Earth answered all the objections against panspermia. Sagan's objection at the Bjurakan conference that radiation would kill microorganisms did not apply to those found in ALH84001. If the microorganisms were deep enough, the radiation would not kill them. However, another problem remains. If we look at the sky every now and then observe light trails produced by meteors that, upon entering the Earth's atmosphere, heat up and burn. The same thing must have happened to AlH84001 when it was thrown into space from Mars. In fact, bubbles formed in ALH84001 when it was heated to high temperatures, trapping atmospheric gases. Upon arrival on Earth, it was again heated to high temperatures, so the high temperature would have killed all life. Those found by meteorite scientists are only traces of previous life. This problem can be solved only if the meteorite is made of heat-resistant material or if its size is much larger than ALH84001. However, ALH84001 had not ceased to amaze us. In 1998, A.J.T. Jull used the radiocarbon technique on meteorites. This technique was invented in the 1940s by W.F. Libby and perfected in 1993 by Minze Stuiver and colleagues, who calibrated the technique using trees, i.e., the dating technique based on counting growth rings in the trunk. The radiocarbon method is based on the following principle. The atmosphere is constantly bombarded by cosmic rays that shatter nitrogen atoms in the upper atmosphere and form radioactive carbon atoms, carbon-14. Carbon-14 reaches the soil and is absorbed by living beings who take it in with food. Together with carbon-14, they absorb carbon-12 and the isotope carbon-13. When an animal dies, it stops taking in carbon, and carbon-14 continues to decay, giving rise to non-radioactive elements. After 5730 years (half-life of carbon 14), half of the radioactivity present in the animal when it is alive will have vanished. After another 5730 years, a quarter of the original radioactivity will remain, etc. In this way, the time of death of the animal can be dated. Jull applied the technique to fragments of ALH84001. He burned fragments of the meteorite and obtained the results of the combustion of hydrocarbons in the meteorite, so he isolated the carbon. Jull also isolated the carbon present in the inorganic calcium carbonate that was part of the meteorite. The ratio of carbon-14 to carbon-12 and 13 was identical to that found in terrestrial carbon, but this ratio was different from that of carbonate, which most likely came from Mars. The conclusion was that the organisms in the meteorite had entered it while in Antarctica. It could only be concluded that the meteorite had come from Mars, but nothing conclusive could be said about life on Mars. If this is the case, ALH84001 does not provide information about panspermia. We are sure that there is certainly no shortage of cosmic "messengers" that can carry microorganisms from one solar system to another.

3.3   Cosmic Messengers

Our solar system, in addition to the planets that compose it, is made of smaller objects. In 1801, Valtellina abbot Giuseppe Piazzi, then director of the Palermo Observatory, discovered the dwarf planet Ceres, and a whole belt occupied by asteroids was later discovered nearby. This belt originated during the formation phase of the solar system. The presence of Jupiter in some astronomical units likely disturbed the area, and it was not possible to form a planet such as the others; alternatively, the matter in the area where Ceres is located was too little to form another planet. Asteroids are therefore located between the orbits of Mars and Jupiter, and due to planetary disturbances, they can leave this band to move in the solar system. Beyond Pluto's orbit lies another region made up of icy bodies, the *Kuiper belt*.

In 1950, Jan Oort, to explain the presence of comets still present today, assumed that there was a spherical cloud of comets, now called the *Oort Cloud*, located between 20,000 and 100,000 astronomical units (Fig. 3.2).

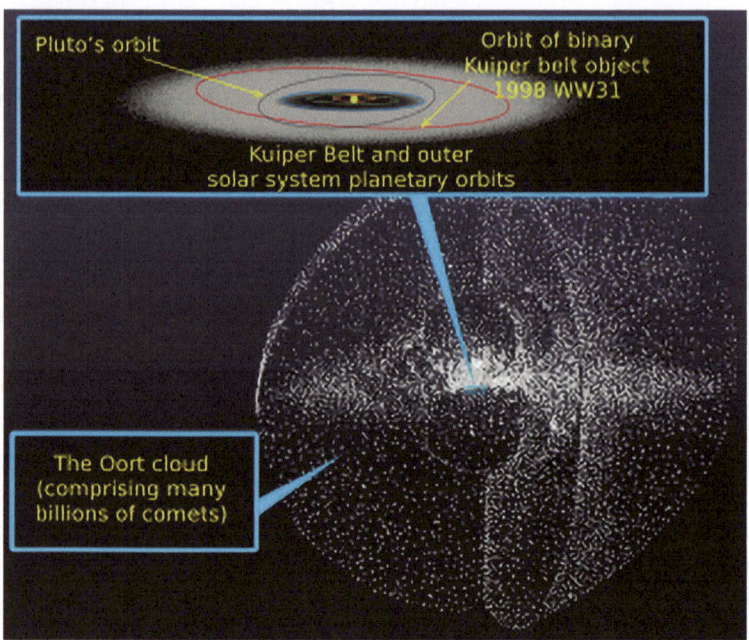

Fig. 3.2 Kuiper belt and Oort's cloud. (Credit: NASA)

Comets that pass close to the Sun are destroyed after a certain number of passes. Therefore, if the comets had all originated at the origin of the solar system, they should no longer exist today, but this is not the case. According to this theory, the Oort cloud would contain millions of comet nuclei, which would be stable because solar radiation is too weak to have an effect at those distances. The cloud provides a continuous supply of new comets, which replace the destroyed ones. The theory seems to be confirmed by subsequent observations, which show us how comets come from every direction, with spherical symmetry. The cloud originated from the gravitational force of giant planets, such as Jupiter, which hurled fragments of matter away from the center of the solar system. The comets of the Oort Cloud are more or less stationary objects, as they are not affected by the action of the planets, but gravitational perturbations can cause them to move toward the center of the solar system. Other stars are also supposed to have Oort clouds. When they approach our Sun, there may be an exchange of comets between the Oort Clouds. It is also believed that wandering comets exist. Between us and the nearest star, there could be 50 billion, and in our galaxy, there could be approximately 10^{24} (1 followed by 24 zeros). Therefore, there is no shortage of couriers to transport microorganisms within our galaxy, and the panspermia theory could make sense. Is there evidence in favor of this theory? Recent studies conducted in India have shown that at altitudes greater than 40 km in the atmosphere, where mixing with the lower layers of the atmosphere is unlikely, bacteria are present. In 2009, bacteria adapted to the stratosphere were also found. Other evidence in favor of panspermia is the rapid appearance of life on Earth, as already mentioned. Fossil stromatolites have been found, i.e., bioconstructed sedimentary structures, due to the activity of bacteria, such as *cyanobacteria*, dating back 3.8 billion years, only 500 million years after the formation of the oldest known rocks. With respect to the possibility of resistance to high temperatures and adverse environmental conditions, bacteria and organisms have been found in *abyssal fumaroles*, which we will discuss in the next chapter. *Extremophilic bacteria* live at temperatures above 100 °C, whereas others live in very caustic environments and are able to withstand enormous pressures and lethal radiation. Bacteria are present in underground lakes and inside rocks. From ice cores in Antarctica, ice cores taken under a kilometer of surface have been found to show how bacteria can survive on icy bodies such as comets. On February 1st, 2003, *Space Shuttle Columbia* disintegrated into the atmosphere. A sample of approximately one hundred worms survived the accident by landing from 63 km inside a 4 kg container, and a sample of moss was also undamaged. These examples support the theory that life can survive after a journey through the atmosphere. The existence of meteorites from Mars and the Moon on Earth suggests that the transfer of material from other planets occurs regularly. Finally, *glycine* (an amino acid) formed spontaneously in the interstellar clouds. A fundamental thing is that if the panspermia theory is correct, life in the Universe should have similar biochemistry. Panspermia theory does not explain the origin of life but moves it further back in space and time. The time factor could, however, favor Crick and Orgel's point of view on the lack of time necessary for the formation of life on Earth. Life could have originated long before the formation of the Earth and in several steps. Directed panspermia, to protect and expand life in space, is becoming increasingly possible owing to developments in solar sails, the discovery of extrasolar planets, the discovery of extremophiles and microbial genetic engineering. Therefore, to date, the question of whether we are extraterrestrial cannot be answered, but the discoveries of recent decades tell us that it is a possibility that cannot be ruled out.

4

Life on Earth

Reality is just one of the realizations of the possible
Ilya Prigogine

A walk on our planet shows us how it is pervaded by life, both animal and plant. The plants cover its emerged parts, dressing it like a lush dress.

From the microscopic to the macroscopic, our planet is teeming with life, and we are part of this large family. Life fills every possible niche, as if it had the tendency to occupy all the space, like gases do. One cannot help but be struck by this triumph of the living, by the touching shapes and colors. However, the Earth was not always like this, full of life. Many billions of years ago, there was no form of life that then timidly appeared and advanced from microscopic unicellular forms to generate the complex forms and species that we observe today. It is natural to wonder how life began on our planet. Giving an answer is extremely complex, and even today, we do not have a real answer. Many scientists have attacked the problem from different points of view over the last century, but no one has succeeded in solving it. Despite this, research has made considerable progress, and we hope to be able to provide a solution to this problem in the future. The possible answers that can be given to this question depend on our nature, whether we are believers or not. Believers believe that life was created by a deity. As we know, there are many ideas on the origin of the world linked to divinities. These beliefs have always been proposed as science, without there being any explanation of the creator God. In more recent times, a more cryptic theory has appeared, *intelligent design* that does not deny the reality of evolution but maintains that the extreme complexity of living beings can be explained only by the existence of an intelligence that directed the evolution to follow the paths he followed. Another possibility is that life would never have had an origin; life would always have existed, according to the ideas of the astrophysicist Fred Hoyle, who based them on his *theory of stationary status*, according to which the universe would be eternal. His collaborator Wickramasinghe also supported this thesis and the idea that space was full of life (viruses, bacteria, etc.). With the discovery that the theory of the stationary universe was wrong, the idea that life had always existed also fell into disuse. If life has not always existed, we must start from the premise that the first living forms originated from nonliving material.

As we discussed in the previous chapter, there are those who think that life did not originate on Earth but came from space (panspermia). Finally, most scientists believe that life originated on Earth. Today, we still do not know how things happened, but step by step, we are getting closer to knowing how they could have happened. Despite the progress made in this area, controversy still exists between the idea that life is the result of chance, according to the ideas of Jacques Monod,

> *Chance alone is at the source of every innovation of all creation in the biosphere. Pure chance, absolutely free but blind, at the very root of the stupendous edifice of evolution*

or of a necessity imposed by natural laws, as supported by Ilya Prigogine or Christian de Duve, who had even written an article in the Phylosophical Transaction of the Royal Society entitled *Life as a cosmic imperative,* in which he wrote:

> *The origin of life may have been close to obligatory under the physical-chemical conditions that prevailed at the site of its birth.*

The study of the origin of life is so complex that it requires interdisciplinary work. Nobel Prize winners in physics, chemistry and biology have dealt and continue to address this complex and extraordinarily interesting problem; however, they have not yet arrived at the solution. The reconstruction of the history of life works quite well between the present and the so-called LUCA (*Last Universal Common Ancestor*), which would be a hypothetical common ancestor cell, from which the three domains of life originated: *bacteria, eukaryotes* (including plants, amoebae, fungi, animals and microorganisms) and *archaea* (or ancient bacteria). The most puzzling aspect is the period from lifeless Earth to LUCA (Fig. 4.1). In the middle of this period, the first beings

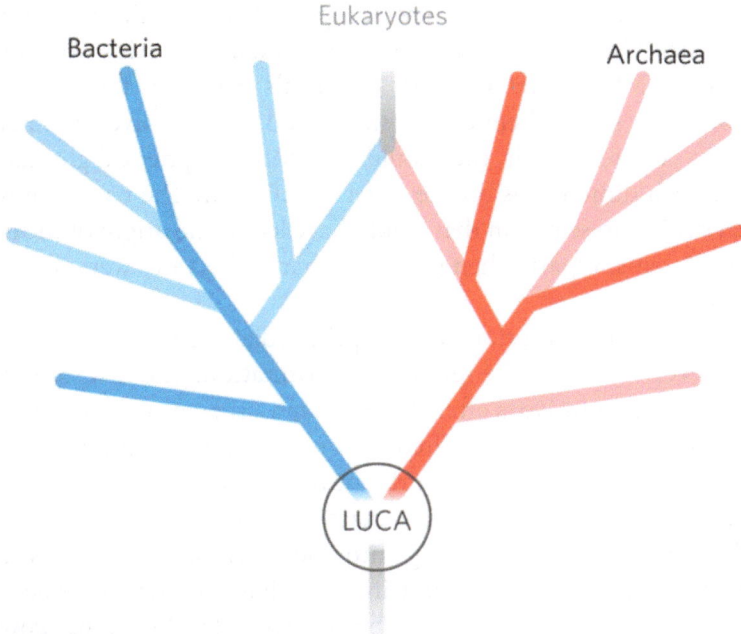

Fig. 4.1 LUCA and the tree of life. (Credits: NASA & WEISS Et al/NATURE MICROBIOLOGY)

(including the ancestors of LUCA) would have formed, preceding LUCA. The ancestors of LUCA gave rise to extinct descendants as well as LUCA.

4.1 The Earth in Its Origins

As Alexander Ivanovich Oparin argued,

> *The origin of life is an inalienable part of the general process of the development of the universe and, in particular, of the development of the Earth.*

For this purpose, it is necessary to discuss how the Earth evolved and the conditions present at the time that life appeared. Oparin imagined the conditions in which the Earth found itself at its formation: a very hot semimolten surface, with a continuous fall of meteorites on it, and a vast range of chemical substances, including those based on carbon (organic reactions). Then, the Earth cooled, and the water vapor condensed into liquid water, resulting in rain. Thus, the oceans were formed, warm and rich in chemical substances.

These may have reacted to form new compounds in the direction of increasing complexity. Oparin suggested that sugars and amino acids were formed in the waters of early Earth. The new chemicals began to form microscopic structures. Since some chemical substances do not dissolve in water, upon coming into contact with it, they could have formed spherical globules called *coacervates* with dimensions on the order of 0.01 cm. The coacervates would have been subject to selection that would have led to the origin of the dynamic and stable systems. Oparin then proposed that coacervates were the ancestors of modern cells.

Returning to the conditions of the primordial Earth, we know that the Earth, like every planet, is the result of star formation. The first stars appeared a few hundred million years after the Big Bang. They were much larger than the current ones, were a few hundred solar masses, and did not have planetary systems since they did not contain the heavy elements necessary for the formation of planets. Given their large mass, primordial stars have a short life and end their existence by passing into the *supernova phase*, in which the heavy elements that have been built in the stellar interior are projected into space, as shown in the famous article by Mr. and Mrs. Burbidge, Fowler and Hoyle. These elements, in turn, can give rise to a second-generation star system, which could already have planets made up of heavy elements. The relative abundance of heavy elements in our solar system suggests that it is a third-generation system. Life therefore appeared only after second- or third-generation planetary systems appeared in the Universe. Estimates of the Earth's age converge at approximately 4.5 billion years, using dating of the oldest meteorites, for example. Our solar system, similar to the other systems, was formed by the collapse of an enormous nebula of gas and dust with a diameter of a few tens of light years, made up mostly of hydrogen. For a collapse to occur, an external disturbance, such as the explosion of a supernova or the passage of a star, is necessary. During the collapse and decrease in size, the density increases with increasing temperature, as the gas is compressed. In the collapse phase, the conservation of *angular momentum* (a typical quantity of rotating bodies) increased the rotation speed of the system. This phenomenon is the same as what happens when a skater moves his arms closer to his body. The result was the formation of a disk-like structure, namely the protoplanetary disc. As the size decreased, the density and temperature increased to the point of triggering hydrogen fusion reactions, forming a star that contained 99% of the mass of the star system. The remaining part was distributed over a disk of gas and dust, the protoplanetary disk. When the disk begins to radiate, its temperature decreases, and small particles form through condensation. They join together to form larger objects that, owing to gravity, attract other matter. This phenomenon, known as *accretion,* gave rise to the formation of many objects of a few kilometers, the *planetesimals,* and the accretion process continued until the *planets formed.* It took a few million years for planetesimals to form and between ten and one hundred million years for planets to form. The newly formed planets were spheres of silicates and metals. Owing to gravity, the denser materials moved toward the center of the planet, and the less dense materials remained in the outer parts. The volatile materials remained outside and in fairly massive planets, such as Earth, where they were captured and formed atmospheres. Thus, *rocky planets* were formed closer to the Sun. Planets at greater distances, for example, 5 astronomical units, continued to grow until they formed a rocky core of approximately 10 solar masses, and at those distances, they captured the large amount of gas that surrounded them to form *gaseous planets.* To the separation between rocky and gaseous giants took an important role the migration of the giant planets near the Sun and the resulting resonance between the proto-Jupiter and proto-Saturn. The formation of the Earth occurred, as with the other planets, by agglomeration. In the period between ten and one hundred million years, a body the size of Mars (called Theia, the name of Selene's mother, which indicates the name of the Moon) collided with the Earth. The large impact hurled a fraction of the Earth's mass into space, which, after subsequent interactions with the Earth, formed the Moon. The event was of considerable importance for the Earth, since our satellite, according to some studies, stabilizes the oscillations of the Earth's axis (other studies, however, deny this) and consequently the seasons. At that time, the Earth was molten and obviously could not support life. The period from the formation of the Earth up to approximately 4 billion years is called the *Hadean eon* from the Greek term Hades, meaning hell, to indicate the inferred conditions existing on the planet at that time. The Earth cooled quickly, so much so that approximately 4.4 billion years ago, a crust probably already existed on which water began to accumulate. Under these conditions, life could theoretically have appeared, but if this happened, it could not survive a period of intense bombardment by meteorites and comets that occurred between 4.1 and 3.8 billion years ago, called *late bombardment.* According to most scholars, life could have appeared and been preserved only 3.8 billion years ago, a time in agreement with some fossil remains, which we will discuss shortly. If this happened, life developed very quickly, a few hundred millions of years after the formation of the Earth. This would indicate that life is a probable process and that extraterrestrial life must therefore be expected to exist. From an astronomical point of view, it is estimated that 3.9 billion years ago, the Sun must have emitted approximately 30% less radiation than it did today; therefore, the Earth should have been completely frozen, a sort of *snowball*. This seems to contradict the *Isua* findings, which we will discuss, which would testify to the existence of liquid water. This contradiction could be overcome by considering that at that time, the atmosphere was such as to give rise to an intense *greenhouse effect*. It is therefore natural to ask what the composition of the Earth's atmosphere was at that time, and there is much controversy on this point. Considering the most accepted ideas, the atmosphere is composed of gases that tend to acquire electrons in reactions, such as hydrogen and helium, and in smaller quantities, such as water, methane, ammonia, nitrogen, etc. Due to solar wind, high temperature, and meteor impacts, lighter gases (hydrogen and helium) are lost from the atmosphere. Methane and ammonia concentrations decreased due to solar radiation. The atmosphere was replaced from that coming from the bowels of the Earth due to the continuous emission of gas from volcanoes, i.e., the *Great Eruption*. As a consequence, the concentration of carbon dioxide increased, with a concentration one hundred to a thousand times higher than the current concentration, together with those of water, nitrogen, ammonia, methane, and carbon monoxide. The early Earth already had water on its surface, as mentioned above. In addition to the water already present, more arrived because of comets coming from the *Kuiper belt* or the *Oort Cloud*. Together with comets, asteroids also fall on Earth in greater numbers than they do today. This brought a large quantity of water, which was added due to volcanic degasification. An observer on Earth would have had a very different view from the current one, with a moon much larger than the current one being at a distance one third of the current one, the seas probably brown–green and a sky red–orange. The tides were much greater than they are now, and the day length was almost half of what it is now. An important question we need to ask is when life appeared on Earth. The oldest fossils are used to date their beginning. Until the mid-1950s, the fossils found served to date the *Cambrian period* (541 million years ago). For the previous periods, there was not much information. After 1954, fossils were found in *Gunflint*, Canada, dating back approximately 1.9 billion years. Other fossils, called Isua spheres, were subsequently found in Isua, Greenland, with an age of approximately 3.8 billion years. William Schopf, a paleontologist, defined biogenicity criteria to prevent formations of nonbiological origin from being considered fossils. With these criteria, the spheres of Isua were introduced into the unidentified category, whereas the others were discarded. The oldest fossils identified were those from *Marble Bar*, Australia, aged 3.5 billion years, identified as cyanobacteria. The latter are green bacteria because they use photosynthesis. The biogenicity of Schopf microfossils is not completely certain. As shown by Juan Manuel Garcia Ruiz, biological structures can be similar to biomorphs of inorganic origin. Therefore, Schopf's structures are microfossils on which there are doubts, pseudo microfossils. Of particular importance are the structures called *stromatolites* (from the Greek "covered by stone"). Stromatolites are generated by the action of microorganisms, particularly *cyanobacteria*. They have disparate shapes ranging from flat to column shaped. More recently, in 2011, remains were found by Brasier near the structures identified by Schopf, which have a greater guarantee of being authentic and which date back approximately 3.4 billion years ago. In addition to morphological fossils, there are chemical fossils. The oldest layers are those of Isua, dating back 3.8 billion years. There are also those from *Akilia*, in Greenland, dating back 3.85 billion years ago, but there are considerable doubts about their biogenicity. Ultimately, from the fossils, we conclude that life on Earth must have already existed 3.4 billion years ago, and perhaps even earlier, 3.8 billion years ago. In other words, life would have already been present 700 million years after its formation.

4.2   The Little Warm Pond

Given a definition of what life is, it is not trivial and even less trivial to talk about its origin. Despite this, Darwin had some ideas about its origin that he did not publish but that he spoke about in his private correspondence. In a letter from 1871 addressed to his friend Hooker, Darwin spoke of the origin of life, starting from chemical processes fueled by energy sources. In the letter, he spoke of a "little warm pond" as a possible primordial soup in which the first living organisms would have formed. In his words,

> However, if (and what a big if) we could conceive in some warm little pond with all sorts of ammonia and phosphoric salts, light, heat, electricity, etc., that a protein compound was chemically formed and ready to undergo more complex changes.

and a few years later, in 1882, to Daniel Mackintosh:

> Although there is still no evidence in favor of the hypothesis that a living being developed from inorganic matter, I cannot help but believe in the possibility that this will one day be proven.

Only in the 1920s was the problem of the origin of life taken up again by the Russian biochemist Alexander Ivanovich Oparin and the English geneticist John Haldane. Both conceived the idea of *chemical evolution*, that is, the idea that, in primordial seas, organic soup, which increases in complexity, would lead to the formation of simple cells, the point of origin of all living beings. Owing to the poor development of analytical chemistry, for many years, there were no developments or experimental ideas to verify the ideas of Oparin and Haldane. The English version of Oparin's second book was read by Harold Urey, the Nobel Prize winner for chemistry, in 1934. In a seminar held at the University of Chicago on the origin of the solar system and, in particular, on experiments for the formation of organic compounds, it was a recent graduate in Chemistry, Stanley Lloyd Miller. Some time later, Miller appeared in Urey's studio, proposing the implementation of some experiments. Urey agreed to perform the experiments, but if there were no positive results in 6 months, they would change Miller's thesis. Together, they built a device containing liquid water and gases: hydrogen, ammonia, and methane. At that time, this was the idea about the constitution of the Earth's atmosphere. The water was boiled in a pipette at the bottom and then cooled to form what was rain in the primordial oceans. Electric discharges of 6000 volts were produced in the pipe above, representing the lightning of the primordial Earth. After a few days, the color of the water changed, becoming reddish-brown. The analysis revealed that various organic compounds had formed, including amino acids (*glycine* and *alanine*), which, as already mentioned, are the "building blocks" that form proteins. The results of the experiment provide considerable support for the idea of chemical evolution. The experiment generated great expectations and provided a strong impetus to carry out other experiments. Importantly, the atmosphere used by Urey and Miller actually differed from what was thought to be the atmosphere of early Earth. By repeating the Urey–Miller experiment with an atmosphere rich in carbon dioxide, which is more similar to that of the primordial Earth, the reaction yield was much lower, i.e., not all the substances in the Urey–Miller experiment were formed.

It is assumed that the contribution of organic matter from space on meteorites and comets could compensate for the loss. Nonetheless, the experiment involved prebiotic chemistry, spurring the performance of a multitude of other experiments. Notably, the results obtained by the Spaniard Joan Orò. Until the late 1950s, the nitrogenous bases of the nucleic acids that we discussed in Chap. 2 had not been identified. In 1959, Orò conducted an experiment starting from hydrocyanic acid, obtaining a base of nucleic acids, adenine (A), and together with Miller, they found guanine (G), while all the other bases were found by other researchers. Thousands of early Earth simulation experiments were performed using different forms of energy (ultraviolet radiation, visible radiation, heat, radioactivity, electrical discharges) in different environments (aquatic, gaseous, and atmosphere–water interfaces). Most of the experiments highlighted the fundamental role of hydrogen cyanide and formaldehyde. The first is a toxic element due to the presence of cyanide ions, and the second, when put in water, gives rise to formalin, a classic preservative for corpses. It is truly paradoxical that two such elements could have been key elements in the birth of life. In 1969, the *Murchinson meteorite* fell, which takes its name from the place where it was found and which is located in Australia. It is a carbonaceous chondrite containing 14,000 different organic compounds, including 70 amino acids, but millions of them are estimated to exist. In addition to amino acids, the bases of nucleic acids have been identified. These substances are not due to contamination, as proven by the presence of organic compounds that are not found on Earth. The composition is similar to that of the results of the Urey–Miller experiment. This finding is of considerable importance because even if the reactions that gave rise to them did not occur in an environment similar to that of the Urey–Miller experiment, the materials important for life revealed that these materials can be formed easily in a natural way. The contribution of extraterrestrial material could play a fundamental role in the birth of life. A few years ago, the study of comet *67 P/Churymov-Gerasimenko* revealed the presence of hydrogen cyanide, nitrogenous compounds, aldehydes, and alcohols. Therefore, comets may also have brought a certain amount of probiotic molecules to early Earth.

4.3 The World of RNA

Even if all the basic components of living beings are present on Earth, the path to forming the first organisms, particularly the LUCA, is long. Life is supported by the work of *enzymes*, the proteins that catalyze or accelerate the network of biochemical reactions that make up *metabolism*. Proteins are synthesized from 20 amino acids. The latter join together to form *polymers*, long chains that follow an order determined by the nucleotide sequences of the DNA that encode them, giving rise to genes. For the functioning of an ancestral organism, enzymes are necessary. They are synthesized from the information contained in DNA, which, as we observed in Chap. 2, are transcribed into RNAs. At the same time, enzymes are needed for DNA to duplicate itself and provide information. That is, we are faced with the chicken and egg problems in the case of the origin of life. When it appeared to be we were on a dead end road, a fundamental discovery was made. Thomas R. Cech, in 1982, and Sidney Altman, in 1983, noted that some RNAs have catalytic abilities; that is, they can function as enzymes, an ability that was thought to be exclusive to proteins. These RNAs that can function as enzymes are called *ribozymes*. The consequence of this discovery was that it was no longer necessary to hypothesize the presence of the proteins and DNA that served for their coding. In this RNA world, RNAs are capable of performing almost all functions important to life: catalyzing biochemical reactions and carrying genetic information. However, in experiments related to early and prebiotic Earth, the difficulties in generating an RNA led to the belief that there were other polymers with capabilities similar to those of RNA, the so-called pre-RNAs. In early studies of the RNA world, the existence of membranes that protect RNAs was not taken into account, but it was realized that RNAs were unlikely to be found in free solutions. Subsequent studies on RNA have led us to believe that, from the beginning, the membranes involved RNAs forming precells. These processes originated in surface waters. However, there is not unanimity that this environment is the most suitable for the birth of life.

4.4   The World of Iron-Sulfur

Günter Wächtershäuser described how life could have originated on the ocean floor, near underwater thermal springs called *black smokers* (Fig. 4.2). Ecosystems are rich in organisms such as crustaceans and worms, whose existence depends on the flow of heat and materials from the Earth's interior. According to Wächtershäuser, the first organisms were not made of cells; they had no enzymes, DNA or RNA. However, they would have had a certain metabolism and a capacity for evolution. Wächtershäuser thought of a volcano from which a flow of hot water, rich in volcanic gases such as ammonia and traces of volcanic minerals, flows. Water flowing over rocks produces chemical reactions, resulting in metabolic cycles. Wächtershäuser's model was developed by him in the 1980s and 1990s, in a very detailed manner, outlining which minerals were in play and which chemical cycles took place. Organic matter is produced by the transformation of carbon monoxide and carbon dioxide starting from compounds of sulphur, iron and hydrogen because of the energy generated by the formation of pyrite, a compound of iron and sulphur. Wächtershäuser called his hypothesis the *iron–sulfur world*. Hydrothermal vents with temperatures below 150 °C, containing pyrite, were discovered in the 1980s by geologist Mike Russel. Russel suggested that thermal vents in the deep sea, which are warm enough to allow pyrite structures to form, harbor Wächtershäuser organisms. Starting from the ideas of Peter Mitchell, Russell concluded that the ideal place for the formation of life is hydrothermal vents with alkaline water.

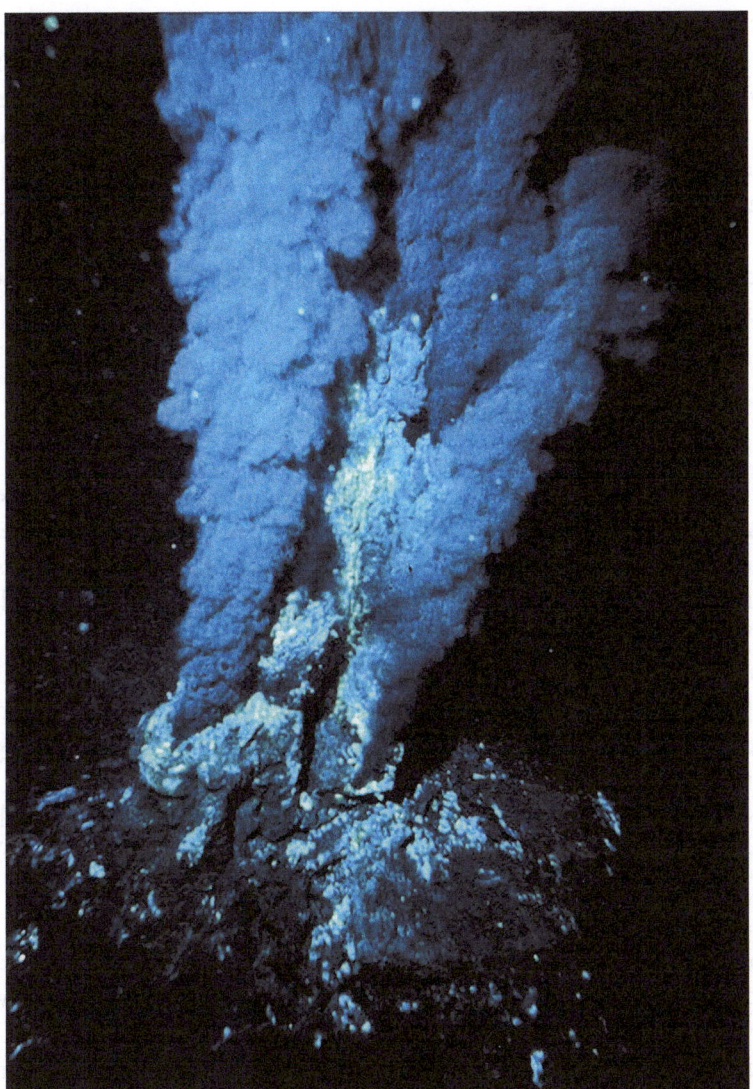

Fig. 4.2   Black smokers. (Credit: NOAA Photo Library)

The first *alkaline hydrothermal vents* with temperatures between 40 and 75 °C and slightly alkaline water were discovered by Deborah Kelley in the Atlantic in a location called Lost city. These *hydrothermal vents* were perfect for Russell's ideas, and he convinced himself that these vents were actually the places where life was born. After life harnesses the chemical energy of vent water, it begins producing molecules such as RNA. With the formation of the membrane, a true cell is formed, which then moves from the porous rock to the open sea. Jack

Szostack is one of the researcher who does not believe in the origin of life in the deep seas, since, according to him, there has never been any evidence of prebiotically plausible chemical reactions in the quoted environment. Thus, life would have arisen in surface waters according to the world of the RNA model or in deep seas according to the model of Wächtershäuser. No firm conclusion has been reached to date, and there is a hot controversy among researchers who believe that life originated in the two aforementioned environments. In 2019, Laura Barge and other reserachers of NASA and CalTech recreated in laboratory the hydrothermal vents to understand if life could really be born there.

4.5   The World of Lipids

In current biology, *metabolism, genetics, and cellularity* are closely linked, and a living organism consists of all three. Given this difficulty, the researchers chose to attempt to obtain the three characteristics separately. In the first studies of the RNA world, membranes were dispensed, even if it is clear that it is unlikely that RNAs would be found in solution without protection. In the hypothesis of Wächtershäuser, the membranes appeared late. Cellularity depends on membranes, which are not simple semipermeable barriers but also have important metabolic capabilities and are fundamental for energy generation. In the last decade, some discoveries have led us to consider a new approach that can simultaneously realize the three functions on which life is based—genetics, metabolism and cellularity—to create an entire cell from scratch. Various attempts have led to the simultaneous acquisition of two characteristics: cellularity and genetics. Michael Russel highlighted the importance of iron-sulfur membranes.

In today's cells, membranes are made up of lipids and proteins. Owing to lipids, which are polar and nonpolar,[1] the vesicles may close. In 2015, Deamer showed how lipid vesicles can give rise to the formation of chains of nucleotides through cycles of water intake and loss. Vesicles can engulf RNAs along with water. These findings led to the idea that a *lipid world* existed before the RNA world. In 2013, Orgel and one of his students, Kataryna Adamala, managed to carry out RNA replication inside fatty acid vesicles, whereas Jack Szostak's team managed to construct protocells capable of absorbing molecules from the outside and retaining their genes. In this way, the protocells can grow and divide, and the RNA can replicate inside. At this point, one last step remains to be able to integrate the third function: metabolism. Succeeding in doing this would provide a unified approach to the origin of life.

---

[1] A molecule that presents a partial charge positive on one side and a negative partial charge on the opposite side of it is called polar. Molecules that do not exhibit the phenomenon of polarity are called apolar.

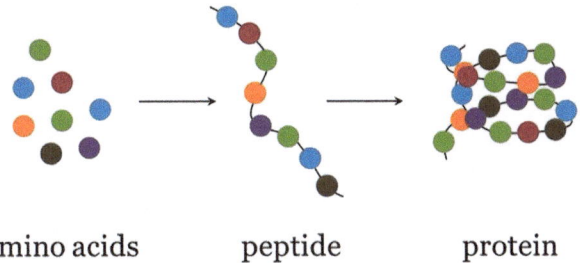

amino acids   peptide   protein

Fig. 4.3 Aminoacids and proteins

Ultimately, there is still no complete recipe to explain the origin of life on Earth. Summarizing the previous theories, we can say that life could have formed in shallow waters, where life is linked to sunlight, with mild temperatures, and the element that would have led to life is RNA. The other possibility is that life originated in deep waters, where it depends on local thermal differences determined by chemical reactions. In such places, temperatures are high, and the basic element is metabolized. Most recent studies have attempted to explore three fundamental aspects of life: reproduction, metabolism, and cells. That is, we search for a mechanism in which all three aspects are at work. **The interested reader can have a look to Appendix B to have a deeper insight in the origin of life on earth.** Given the difficulties of reproducing life on Earth, there are scientists who continue to think that life was born outside the Earth and arrived there or that at least some of the processes that led to life were started in space (the panspermia we discussed in Chap. 3). In 2019, Nature Astronomy published an article by T. K. Henning and S. A. Krasnokutski in which the conditions were included in the interstellar clouds. In a very high vacuum chamber at −263 degrees centigrade, Krasnokutski and colleagues placed some components of interstellar clouds, such as carbon, carbon monoxide and ammonia. In this way, they obtained peptides, which are the building blocks of proteins (Fig. 4.3).

Not everything is clarified because the passage between this phase of study and the subsequent phase that leads to the formation of living beings is missing. Furthermore, we need to know whether the interstellar peptides have stood the test of space travel and subsequent impact with Earth.

However, other studies showed that purines and pyrimidines also are produced in experiments simulating the space environment.

The panspermia hypothesis rears its head again in our history. In conclusion, we know that life exists on Earth, but we are not able to reproduce the steps it followed to arise on our planet. Even if it had been born elsewhere in the Universe and had been transported to Earth, we would always be left with the dilemma of its formation. Are there other planets or satellites in our solar system that host life, including microbial life? We address this in the next chapter.

5

Life in the Wanderers of Sky

*Extraterrestrial take me away, I want a star that is all mine, extraterrestrial come
and find me, I want a planet to start again on*
Eugenio Finardi

I do not know if you have dedicated yourself to observing the planets in the
sky. First, they have static light, unlike stars, whose light sparks.[1] Another
characteristic of them, if they have the patience to observe them for several
months, as the ancients did, is that they show a strange motion, different from
that of the stars. Typically, the planets move eastward with respect to the stars
in motion called *direct motion*. Sometimes, they reverse their motion and
move westward. This is their *retrograde motion*. After a period of retrograde
motion, they change the direction of motion again and continue moving in
the original direction. The time that passes between two successive retrograda-
tions depends on the planet. For example, for Mercury, this happens every
116 days, and the planet moves retrogradely for approximately 21 days; for
Jupiter, it moves every 399 days, and the retrograde motion lasts 121 days.
The retrograde motion of Mars particularly disconcerted ancient astronomers
because, in a geocentric system, Mars orbit, during retrogradation, appears to
pierce that of the Sun. Because of their strange motion, the Greeks gave the
planets the name of *plànētes asteres*, "wandering stars". The Greeks knew 6 of
the planets we know: Mercury, the messenger of the gods, Venus, the goddess

---

[1] Scintillation is due to the bending of light in different layers of the atmosphere. The planets do not
twinkle because they are closer than the stars and the front of the wave that reaches us is wider than that
due to the stars, therefore the light from the planets passes through more layers in the atmosphere and the
effect of deviation of the light compensates and we do not see them sparkle.

A. Del Popolo, *Extraterrestrial Life*, https://doi.org/10.1007/978-3-031-83497-4_5

of beauty, Mars, the god of war, Jupiter, the father of all gods, Saturn, the lord of time. Even the sun and the moon were planets for them. Over millennia, their number had increased to 9, with the addition of Uranus, Neptune, and Pluto. After 2000, objects located beyond Neptune, i.e., *trans-Neptunian objects, were discovered*: in 2003, Haumea was slightly smaller than Ceres, Sedna had a diameter of almost two thousand kilometers, and in 2005, the discovery of Eris with dimensions and masses similar to those of Pluto was announced. After these discoveries, Pluto's status as a planet was rethought. One fine day in 2006, Pluto was demoted to a *dwarf planet*; therefore, today, the number of planets is 8. A definition of a planet was also introduced on August 24, 2006, by the *International Astronomical Union*, according to which a planet is a celestial body that orbits around a star and does not produce energy through nuclear fusion; its mass is large enough to give it a spheroidal shape, and owing to its gravitational force, it manages to keep the orbital belt free from other bodies of comparable or larger dimensions. Pluto does not satisfy the last condition. Now you might be wondering why I'm talking about planets given that our discussion is related to life and, in particular, to extraterrestrial life, given that you will surely have heard that in our solar system, there is life only on Earth. Well, to be honest, we're not truly sure. If you remember, when talking about ALH84001, we said that perhaps it brought life from Mars to Earth. We have seen that there are some experiments that disagree on this, but we cannot be 100% sure that there was no Martian life in ALH84001. We will also discuss experiments carried out on Martian soil, which still spark debates about whether there is life on the red planet. Furthermore, as we will see, it is possible that there is life in some satellites of the gas giant planets (Jupiter, Saturn, etc.). The objective of this chapter is precisely to discuss whether life can exist in our solar system. Microbial life, obviously. We do not expect to find humanoid Martians. These traces can be revealed directly or through chemical or energetic manifestations. In other words, it is necessary to understand whether probiotic molecules are present or can be formed and whether there is a source of energy and an adequate liquid medium. Light energy is the most effective energy source for biological processes, and the sun provides this energy to all bodies in the solar system. A favorable chemical environment presupposes the existence of carbon and organic molecules, and the presence of compounds of oxygen, nitrogen, sulfur, and phosphorus is also important. Currently, carbon, hydrogen, oxygen and nitrogen are among the most abundant in the Universe. The probiotic compounds could come, as on Earth, from space or perhaps form on site. It is very likely that all the bodies in the solar system have a minimum of these probiotic compounds. For life to arise from these molecules, a liquid environment on the surface or below it is essential. The habitability of a "world" is closely linked to the presence of a liquid such as water or others such as ammonia, simple hydrocarbons, etc. There are several places in the solar system that meet this requirement. In 2001, the astrobiologist Schulze-Makuch and others proposed an index to evaluate the habitability of a planet and its moons, the *PHI index* (planetary habitability index). This parameter is based on the variety of chemical elements and physical characteristics, such as the presence of a solid substrate, the availability of energy, the presence of liquids and a favorable chemical environment. Earth has a PHI of 0.96, followed immediately by Titan with 0.64, Mars (0.59), and Europe (0.49). Another useful index is the Earth similarity index (ESI). By definition, the Earth has an ESI equal to 1, and then Mars has 0.70, Mercury 0.60, the Moon 0.56 and Venus 0.44. This index is less useful than the PHI for the bodies of our solar system, whereas it is more relevant for extrasolar planets for which there is not much data on habitability.

5.1   The Wastelands

Where the sun beats, And the dead tree gives no shelter, the cricket no relief, And the dry stone no sound of water.
  Thomas Stearns Elliot

Probes have been sent to some planets and satellites in our solar system. Starting with Mercury, not in chronological order, Mariner 10 was launched in 1973 with the goal of flying over and studying Mercury. The observations immediately revealed, as expected, that it was a barren desert.

Mercury is slightly larger than the Moon, as the Moon has no atmosphere or extreme temperatures: approximately 400 °C during the day and −180 °C at night. A place absolutely hostile to life.

Venus is a planet slightly smaller than Earth with scarce quantities of water. Distributing it uniformly would result in a layer 3 cm deep, compared with 3 km deep on Earth. The average temperature is on the order of 464 °C, which is even higher than that of mercury and 90 times higher than that of Earth. These high temperatures are due to a notable *greenhouse effect*. The atmosphere contained 96% carbon dioxide and 3.5% nitrogen. Since Venus is closer to the Sun, it receives more heat, the water has evaporated much more than it has on Earth, and the greenhouse effect has significantly increased the soil temperature, which has produced a further evaporation and an increase in water in the atmosphere, which in turn has increased the greenhouse effect. In the atmosphere, ultraviolet radiation dissolves water into hydrogen, which disperses into space, and into oxygen, which reacts with compounds in the crust. Since the water did not remain on the surface, carbon dioxide did not bind to the rocks and was concentrated in the atmosphere, increasing the greenhouse effect. At heights between 50 and 60 km, the temperature in Venus' atmosphere is between −20 and +70 °C, and the pressure is similar to that of the Earth's surface; consequently, this region could be habitable. However, sulfuric acid is present in these areas. *The thermoacidophilic microorganisms* found on Earth can live in this environment. From data from the Venus Express probe launched in 2005, there has been speculation about the presence of oceans on Venus billions of years ago, when the Sun was colder and the habitability zone was narrower. Therefore, life could have formed and then adapted to the area of the atmosphere indicated above. Proof that life could exist on Venus is the discovery of a compound (carbonyl sulfide) that is difficult to reproduce without the presence of life. Another compound, phosphine, was discovered in the cloud of Venus 4 years ago. The discovery faced controversy, earning rebukes in subsequent observations that failed to match its findings. The same team that discovered the compound has come back with more observations. The data, the researchers say, contains even stronger proof that phosphine is present in the clouds of Venus. Phosphine is potentially a biomarker.

Mars, the fourth planet, now known as the red planet, has a situation related to the presence of water and life that is slightly more positive than Venus. As we observed in the introduction, it is the object of the solar system that has most stimulated the imagination of the existence of extraterrestrial civilizations, the famous Martians. As already reported, in the introduction, the images of Mariner 4 in 1965 were a blow to the heart of those who believed that Mars hosted life. However, until 1976, when famous experiments involving Viking probes were carried out, people continued to hope that primitive life forms existed. This mission was followed by many others, but it is still not 100% clear whether Mars hosts microbial life or whether it has hosted it just in the past. Mars does not have a dense atmosphere because, having a mass ten times less than Earth's mass and a gravity of 38% of Earth's mass, it cannot retain it. Compared with Earth, receives half the total amount of sunlight and therefore has low temperatures, at −46 °C on average. This issue, the absence of a dense atmosphere and the permanent absence of liquid water make the planet unsuitable for life. The planet does not have an ozone layer, and this, together with other compounds present on it, can destroy organic matter. On Mars, channel-shaped structures are observed, which suggests that there were rivers in the past, and photographs from the *Mars Global Surveyor* have shown the effects of water flows in recent years. Images from the *Mars Reconaissance Orbiter* revealed that there would be saltwater on the surface in the summer months. Some studies conclude that approximately 3.8 billion years ago, there were oceans on Mars that then disappeared. The reason is related to carbon dioxide. At that remote time, there must have been a large amount of carbon dioxide on Mars from volcanic activity. With the presence of carbon dioxide, the temperature must have been higher than it is today, with an average above 0 °C. As time passed, carbon dioxide was partly reabsorbed into the rocks, and furthermore, with the cooling of the internal part of the planet, volcanism continued to decline, providing increasingly less carbon dioxide. For 3 billion years, ultraviolet radiation has split carbon dioxide into carbon and oxygen, which have been dragged into space by solar wind. This loss is linked to the absence of the magnetic field that forms a magnetosphere that blocks the solar wind. The absence of the magnetic field is probably due to the internal cooling of the planet, if it is formed with a mechanism similar to that of the dynamo. The remaining carbon dioxide condensed in the polar ice caps. Water would also have suffered a similar end to that of carbon dioxide. Water was probably present on Mars for 2.3 billion years. When it disappeared, if there was life on the planet, it disappeared with the water, and today, we could find fossil remains, as in the case of ALH84001 (but as mentioned, it is not clear whether the remains in the meteorite were of Martian life). Therefore, there was probably life on Mars in the distant past, but does it still exist? If it exists, it is more likely to be found in red sedimentary rocks or below the surface a few centimeters or hundreds of meters. The belief that life still exists on Mars today is linked to the presence of atmospheric methane and data from Viking probes. In 1976, Viking 1 and 2 landed on Mars. In addition to providing photos of the planet, analyses of Martian soil were carried out. No organic matter was found in the analysis, which is unexpected because organic matter literally falls from space. The biological experiments that were carried out focused on *degradative* or *biosynthetic metabolism*. What does it mean? In degradative metabolism, organic nutrient molecules, such as carbohydrates, lipids and proteins from the extracellular environment or from reserves accumulated in the cell, are degraded by successive step reactions into simpler and lower-molecular-weight final products, such as lactic acid, carbon dioxide and ammonia. In biosynthetic metabolism, the opposite occurs: basic molecules are used to form large cellular macromolecular components, such as proteins and nucleic acids. These experiments produced partially positive results. In particular, in the experiments in which radioactive carbon-14 was supplied, carbon dioxide was obtained. Despite this, NASA's conclusions were that there is no trace of life on Mars, a conclusion linked especially to the fact that no organic matter was found. The engineer behind the experiment, Gilbert V. Levin, continues to maintain that NASA's conclusions were incorrect and that the result obtained was due to biological activity. It was recently established that the organic matter present on Mars was destroyed by particular salts (*perchlorates*) that were found on the same soil by the *Phoenix probe* in 2008. High temperatures were used in the tests, and at these temperatures, those salts destroy the organic matter. These results lead to a reinterpretation of the data, giving credence to researchers such as Schulze-Makuch and Joop Houtkooper, according to whom the results have a more logical explanation from a biological point of view and that there was probably life in the Martian samples. In 2015, Javier Martin-Torres, using data from the *Curiosity rover*, revealed how the quoted salts produce macroscopic effects on the Martian surface. The Viking experiments were designed for metabolic processes typical of terrestrial microorganisms. At that time, Sagan and others reported that even if the results were negative, they would have excluded only a subset of Martian microorganisms. Furthermore, the experiments could not exclude the presence of life in other parts of the planet or underground. In 2004, the *Mars Express probe* detected methane in the Martian atmosphere. Subsequent analyses in 2012 by the *Curiosity rover* yielded negative results, but in 2014, they were found again. Such methane can be produced by volcanism or by life. Much of the Earth's methane is of biological origin. Furthermore, the amount of methane in different regions of Mars changes with season, which could lead one to think that it could be produced by *methanogenic microorganisms*. These bacteria are present on Earth. They use molecular hydrogen as an energy source. They are present in groups of microorganisms in which the fermentation of bacteria and fungi releases hydrogen. Even if the methane is volcanic in origin, it has biological implications because the "geothermal" heat needed to release the methane could keep underground water liquid. The *ExoMars program* consists of two missions: the first includes the launch of the *trace gas orbiter*, which took place in 2016, and the second involves carrying the *Rosalind Franklin rover* and will be launched in 2028. This program aims to clarify the question of life on Mars.

Between Mars and Jupiter, as already mentioned, there is a band of small rocky bodies, the asteroid belt.

Ceres, which is 950 km in diameter, is the largest. In 2014, the Herschel Space Observatory confirmed that Ceres contains water and expels up to 6 kg of vapor per second from the atmosphere. NASA's *Dawn probe* has been rotating around Ceres since 2015 to obtain more data on its composition, possibilities and habitability. The mission ended in 2018. From the data, Ceres is thought to have formed in a cold, wet situation and therefore may have water underground. Under 100 km of ice, there could be a rocky core and a salty ocean in the middle. Life forms exist in underwater hydrothermal vents on Earth, but on Ceres, there is probably not enough internal heat. However, carbon and hydrogen molecules that can form the basic compounds of life have been discovered. The Chinese are expected to launch a probe to Ceres this decade to further deepen Dawn's studies.

Pluto was downgraded to a dwarf planet, such as Ceres, in 2006. It was visited by the *New Horizons* mission. It has abundant frozen water and a weak atmosphere composed of methane and nitrogen. It contains organic compounds, but it is unlikely that life developed on it.

5.2   Giant Worlds

After the asteroid belt, the giant region of the solar system begins.

Jupiter is the first giant planet, with a mass more than three hundred times greater than that of Earth. It is a failed star. If it had a mass 13 times greater, according to the latest estimates, it would have been a star, a *brown dwarf*. In brown dwarfs, deuterium fusion reactions are triggered, but not those of hydrogen, and above 65 Jupiter masses, lithium fusion occurs. For hydrogen fusion to occur, a mass greater than approximately 80 Jupiter mass is needed. Jupiter has a very large atmosphere made up of hydrogen and helium, and beneath it, a rocky core of 10 Earth masses with enormous temperatures, 20,000 °C and pressures 500 times that at sea level on Earth. The external temperature, however, is very low, approximately −110 °C. The predominant form of carbon is methane, and acetylene, ethane, and hydrocyanic acid, which are formed as in a large prebiotic chemistry experiment, use internal heat due to their formation and enormous and powerful electric discharges. In this enormous laboratory, it is very unlikely that the elements formed can go beyond giving rise to complex probiotic elements. The planet has many satellites, 95 of which were discovered by Galileo in 1610: the Galilean satellites Io, Europa, Ganymede and Callisto. Among these varieties, Ganymede is the largest and is larger than Mercury. Callisto is not much smaller (Fig. 5.1).

Io, with more than 300 active volcanoes, is the most geologically active object in the solar system. Extreme geological activity is the result of tidal heating due to friction caused by Jupiter and other Galilean satellites. Many volcanoes produce plumes of sulfur dioxide that rise up to 500 km above their surface. Among the four Galilean satellites, the other three are the most interesting in terms of life. They have a surface made of ice, they do not have an

Fig. 5.1 Jupiter and the four Medici satellites of Jupiter in a photomontage that compares their dimensions. From top: Io, Europa, Ganymede, and Callisto. (Credits: NASA)

atmosphere, and the density is indicative of the fact that, in addition to ice, they are made of internal rocks. There is evidence that all three satellites have oceans between the outer ice crust and the rocky interior. At the distance at which these three satellites are located, there should be no liquid water or oceans beneath the surface. This is possible for the same reason that *Io* is covered in volcanoes: tidal forces. These generate enough heat to melt ice and form oceans of liquid water. In 2023, the *European Space Agency* launched the *JUICE* (Jupiter Icy Moon Explorer) mission. The mission's target is Jupiter and its three icy moons. The mission will study the conditions of planet formation and the possibility that life exists, especially on the three moons.

Europe has an icy smooth surface with few craters, with temperatures between −160 °C at the equator and −220 °C at the poles. It was studied by the *Voyager 1* and *Voyager 2* probes and for 8 years by the *Galileo probe*. It has a weak oxygen atmosphere and has *cryovolcanoes*, i.e., volcanoes that emit low-temperature materials such as water, ammonia, methane, etc. The water is thought to have risen from the interior, forming an ice crust between 10 and 30 km thick. Under the ice, there would be a salty ocean with a mass greater than Earth's oceans. The energy that allows liquid water to exist in a satellite thus far from the sun comes from the tidal forces produced by Jupiter. Tidal forces cause constant movement within Europe, which can make the internal ocean warm enough to support life. The Galileo probe identified leaks of carbon dioxide and sulfur in some areas of Europe, both of which are possible signs of volcanism. The heat that volcanoes can generate can rise to the surface and be transported by ocean currents. After Mars, Europe is considered the body of the solar system with the greatest biological interest. Silicates have been found in the crust, and there may be organic matter from asteroids and comets. It is not known whether life exists on Europa; however, the conditions are compatible with life in the oceans. These environments are very similar to the hydrothermal vents present on Earth at deep depths in the ocean and especially similar to Lake Vostok in Antarctica. Lake Vostok has been buried under 4 km of ice for at least 25 million years. The thickness of the ice does not allow any type of photosynthetic process. This environment is an ideal model for determining how a potential biosphere could survive in Europe's oceans. Life in such an ocean might resemble life microbes present on Earth at ocean depths. We discuss the ways of living in hydrothermal vents in Chap. 4. The forms of life on the ocean floor thrive despite the lack of sunlight and constitute a completely independent food chain, the basis of which is a bacterium that derives energy from the oxidation of reactive chemicals, such as hydrogen and hydrogen sulfide, which come from inside the Earth. The existence of life requires only water and energy and does not necessarily depend on the Sun. The analysis carried out with the Galileo probe suggested the presence of organic molecules on Europa. Like what was assumed for early Earth, meteorite impacts were a possible source of organic compounds.

Furthermore, the shock of the impact may have triggered organic synthesis processes. Computer simulations have shown that comet impacts have brought 1–10 billion tons of carbon to Europe, so Europe would have a significant reserve of biogenic elements, with strong implications for the possibility of sustaining life. As mentioned, life could be clustered around hydrothermal vents on the ocean floor, where life forms similar to terrestrial *endoliths* could exist, which they live in very small spaces between one rock and another. Another possibility is that, similar to algae and bacteria in the Earth's polar regions, life could exist close to the lower surface of the ice layer covering the satellite. The existence of some examples of *macrofauna* has also been hypothesized in the presence of significant turnover of the upper ice layer. Richard Greenberg, in 2009, showed that cosmic rays could trigger a process that would mean that Europe's oceans could reach a higher oxygen concentration than Earth's in just a few million years. If life was born as on the Earth, as anaerobic life, oxygen could produce the end of this kind of life but could support large oxygen-using organisms such as fish. In addition to JUICE, more information about Europa will be available from NASA's *Europa Clipper* mission, which is expected to be launched in October 2024.

Ganymede is the largest satellite in the solar system; like Europa does, it has an ice crust and a surface temperature of approximately −173 °C. Its density is 2/3 that of Europa, which suggests that it contains more water than the latter. In 2002, the Galileo probe revealed the presence of a magnetic field probably produced by the movement of saltwater and the metallic nucleus. The Hubble telescope observed a tenuous oxygen atmosphere. Oxygen should be produced as a result of radiation incident on the surface, which causes the splitting of ice molecules present on the surface into hydrogen and oxygen, and this would not be proof of the existence of life on Ganymede. In 1996, ozone was also discovered in the atmosphere. The Hubble Telescope also observed auroras, and from these, it was deduced that the ocean of Ganymede, 100 km deep, probably contains more water than the Earth's oceans do. Unlike that of Europa, the ocean of Ganymede is not located between two icy layers but should have a sandwich structure. The JUICE mission will also study Ganymede.

Callisto is the satellite most heavily cratered in the solar system. Its pockmarked surface overlies a crust 80–150 km thick, whereas at a depth of 50–200 km, it should present a layer of liquid and saltwater that is 10 km thick and heated by radioactivity. This ocean was discovered through studies of the magnetic field around Jupiter and its innermost satellites. Callisto does not participate in the orbital resonance involving the other 3 Galilean satellites; therefore, it does not undergo tidal heating, which gives rise to the endogenous phenomena present on Io and Europa. Without an internal magnetic field and just outside the range of radiation of Jupiter, Callisto is composed, more or less equally, of rock and ice with the lowest density among the Galilean satellites. On its surface, the presence of ice, carbon dioxide, silicates and organic compounds was detected. Callisto is surrounded by a thin atmosphere composed of carbon dioxide and molecular oxygen, as well as an ionosphere. Like Europa and Ganymede, it is thought that extraterrestrial life could exist in a salty ocean beneath Callisto's surface. However, conditions appear to be less favorable on Callisto than on Europa. The main reasons are the lack of contact with rock material and the lower heat flow from Callisto's interior.

Saturn also has many natural satellites, 146. The astrobiological interest of some of its satellites is no less than that of Jupiter's satellites, which we have discussed. For this reason, NASA, ESA, and the *Italian Space Agency* organized a mission, *Cassini-Huygens*, launched in 1997 and completed in 2017, with the task of studying the Saturn system, including its moons and its rings.

Enceladus is the sixth largest satellite of Saturn, with a diameter of 500 km, is icy and is subject to intense tidal forces produced by Saturn and the satellite Dione. These tidal forces heat the satellite interior, causing it to have oceans between layers of ice and rocky cores. In 2005, *Cassini* revealed enormous geysers at the south pole, originating from four fractures approximately 130 km long, 2 km wide and 500 km deep. From these fractures, gas is expelled with water, carbon dioxide, carbon monoxide, methane, ammonia and a cocktail of organic substances of prebiotic interest. The temperature can exceed 90 °C, which indicates that the source below the surface is hot and watery. In the Southern Hemisphere, there should be an internal ocean of saltwater under a 15–25 km ice crust. To establish whether it is possible for life to exist as in terrestrial black fumaroles, in 2015, Cassini crossed one of the geysers 49 km above the ground, where samples were taken for analysis. The probe detected the presence of molecular hydrogen, which is a potential nutrient for microbial life, and evidence of hydrothermal activity in the seabed of Enceladus, a factor favorable to life. Enceladus is the only place in the solar system where hydrothermal activity has been demonstrated; therefore, it is one of the most suitable places to host life (excluding Earth). Given the possible presence of life on the satellite, various missions have been planned (Fig. 5.2).

Titan is the largest satellite of Saturn, with a solid surface and a considerable atmosphere. The satellite was studied by the *Cassini probe,* which arrived at Saturn and its satellites in 2004, followed by the Huygens probe shortly after. Titan's atmosphere has a pressure 50% higher than that of Earth's

Fig. 5.2 Saturn, Titan and Titan's landscapes. (NASA)

atmosphere and is made up of 95% nitrogen, 5% methane and hydrogen. There are clouds of methane, ethane, *polycyclic aromatic hydrocarbons* and other organic complexes. Titan, similar to Earth (which has a water cycle), has a *methane cycle* that includes rain, which, together with ethane and other hydrocarbons, feeds rivers and forms lakes and seas at temperatures of approximately −180 °C. Methane solidifies at approximately −182 °C, which explains the presence of liquid methane. Surface liquid masses were first discovered in solar systems, beyond Earth. In the northern hemisphere, there are large seas. Data from the Huygens-Cassini probe indicate that there may be an underground liquid ocean of water and ammonia at a depth of 100 km. There are dunes of organic matter, especially in the equatorial area. In some experiments, the formation of nitrogenous bases (as we know, the building blocks of DNA and of RNA) and amino acids was observed when energy was applied to a gaseous compound similar to Titan's atmosphere. This was the first observation of the formation of nucleotides and amino acids in the absence of liquid water. The energy sources on Titan are electrons from Saturn's magnetosphere, cosmic rays, and ultraviolet light. Titan resembles a massive Miller–Urey experiment. Since there is no water on the satellite, NASA is not interested in this topic. As we will see in Chap. 9, however, it is theoretically possible that life exists with a solvent other than water, methane, in our case. In 2010, data from the Cassini probe revealed anomalies in the composition of Titan's atmosphere near the moon's surface. In the first study, it was highlighted that hydrogen disappears near the surface. The second highlighted the lack of traces of acetylene, a compound that should be abundant in a methane atmosphere. These results were explained in terms of the presence of life that could have developed even in the absence of water, using hydrocarbons instead. Cellular respiration occurs by absorbing hydrogen instead of oxygen and reacting with acetylene instead of sugar to produce methane instead of carbon dioxide. The consumption of these substances near Titan's surface could, however, be due to inorganic reactions. The presence of cryovolcanoes gives hope that life is based on water.

The Neptune moon Triton also has some biological interest.

Triton has dimensions similar to those of our Moon and appears to have been captured by Neptune. In fact, it rotates in the opposite direction to that of Neptune. It has a weak atmosphere and is geologically active. Beneath the surface of Triton water, there may be an ocean rich in ammonia, liquid nitrogen, and methane. It has been speculated that life based on silicon rather than carbon could exist on Triton. There is also a certain abundance of organic compounds. The possibility of life in Triton's ocean is certainly lower than that in Europe, but it cannot be discounted. However, hypothetical extraterrestrial life on Triton would not be like life on Earth due to extreme temperatures, environmental conditions (nitrogen and methane) and the fact that the moon lies within the dangerous magnetosphere of Neptune, which is harmful to biological life forms.

To rank the most habitable places in the solar system on the basis of the PHI index, Titan is in first place (0.64), followed by Mars (0.56), Europa, Ganymede and Callisto (0.47), Venus (0.37), Enceladus (0.35), Ceres and Triton (0.23) and Pluto (0.22).

Our latest discussions tell us that despite the many efforts made from the 1970s until now, carrying out theoretical studies and sending probes to remote places in our solar system, we have no evidence that life exists in our solar system and that even if it exists, it would be microbial life. We will not find palaces or cities inhabited by other civilizations on any object in the solar system. This is a bit disappointing, but there's not much we can do about it. In our solar system, it seems that we are alone, if we talk about technological civilizations. Despite this disappointment, the search for life in our space surroundings shows no sign of stopping. Several missions are planned to clarify the problem of life on the satellites of the giant planets and on Mars. Why invest money and effort in trying to find at most a few bacteria? There are at least two reasons for this. The first is that man is naturally curious, and this characteristic of ours has pushed us far away: from caves to space. Then, there is a second reason. Finding some form of elementary life in our solar system would make us understand that although we do not know how life arises, as described in Chap. 4, nature is evidently more intelligent than us and has found ways to shorten the birth times of life and that the latter is therefore a natural tendency. If this were the case, the Universe could be teeming with life. This would mean, as already mentioned, following Alexander Ivanovich Oparin

> *The origin of life is an inalienable part of the general process of the development of the Universe and, in particular, of the development of the Earth.*

In other words, the appearance of living organisms is not an accidental event but is implicit in the irreversible processes of systems far from equilibrium, as Ilya Prigogine argued. There is a relationship between spontaneous self-organization and the birth of life. It is as if there is some sort of necessity in the world of nonlife that pushes it in the direction of living. Disorder does not constitute the rule for matter but only an intermediate stage that moves in the direction of creating an ever lower disorder until order and therefore life is achieved.

6

The Universe and Life

The idea that the Earth, our home, an insignificant fragment of an enormous Universe made up of hundreds of billions of galaxies, could be the only place in which life has developed and exists, is mindboggling. As Carl Sagan said,

> *Our planet is a lonely speck in the great enveloping cosmic dark. In our obscurity, in all this vastness, there is no hint that help will come from elsewhere to save us from ourselves.*

Is it truly like this, are we alone in the Universe? Of course, today, we do not have an answer to this question, but if we want to discuss the possibilities of life in the Universe, we need to know it from its origin to today. In this way, we can understand which places are the most likely places where life can appear. We need to understand this information to understand its possible habitable zones.

Today, we know that the Universe originated from the Big Bang 13.8 billion years ago and is experiencing accelerated expansion. Among other things, this expansion does not have the slightest negative influence on life. The hundreds of billions of galaxies that constitute it were formed owing to the action of the force of gravity. The first, according to the latest observations from the *James Webb telescope*, arose a few hundred million years after the Big Bang. The same force of gravity collapses the clouds of gas and dust found inside galaxies and gives rise to stars when four hydrogen nuclei fuse to form helium. Fusion within stars such as the sun does not exceed the formation of carbon. Heavier elements such as oxygen, phosphorus, sulfur and iron, which are also as important as carbon for life, form more massive stars. Stars live as long as they have material for fusion, and the more massive stars consume the material necessary for fusion more quickly and live fewer than smaller stars do. While a star such as the sun can live for 10 billion years, those that have ten times more mass live a few million years and end their life, as we have already seen, with a large explosion, i.e., *supernovae*. They are very useful because they disperse the elements important for life into space. As Sagan said, we are stardust. In Chap. 4, we discuss how planets form. The first planetary systems were lacking in heavy elements, and from this, we can conclude that life as we know it today could not form in them. As discussed in Chap. 4, life is thought to have appeared on second- or third-generation planets, like our own. Second-generation planets probably appeared 2 billion years after the Big Bang. If life exists in the Universe and if it followed paths similar to those followed on Earth, appearing less than a billion years after the planet's formation, this would mean that planets with life could have existed as early as 3 billion years after the Big Bang. In other words, 11 billion years ago, there could have already been planets with bacterial life in the Universe, and approximately 7 billion years ago, complex life, such as human life, could have appeared. In how many places in the Universe could this happen?

6.1　The Cradles of Life

Our universe is very large, and it is even debated whether it is infinite. We can observe objects up to 43.5 billion light years away. In this enormous space, there are hundreds of billions of galaxies, and since each galaxy contains, on average, 100 billion stars, in the observable Universe, there should be something such as 10^{22} (1 followed by 22 zeros) stars (Fig. 6.1).

Fig. 6.1 Spiral galaxy NGC 4414, and galaxy cluster eMACS J1823.1+7822. (Credits: ESA/Hubble & NASA, H. Ebeling)

With this immensity, we find ourselves in the galaxy called the Milky Way, approximately 26,000 light years from its center. The Milky Way is a spiral galaxy with a disk structure with a diameter of approximately 100,000 light years. It completes one rotation in 230 million years. The closest star, Proxima centauri, is 4.2 light years away. If our galaxy could be fit into a 100 m football field, Proxima Centauri would be at 4.2 mm. Surrounding the galaxy is a group of other galaxies that form the *local group*. This group, together with other groups, is part of a cluster of galaxies, the *Virgo cluster*. In turn, the clusters, which can contain tens of thousands of galaxies and have dimensions equal to 30 million light years, form *superclusters*. There are also other structures in the Universe: almost spherical areas as large as clusters with very little galaxy content, called *voids*, are delimited by filaments of galaxies. A legitimate question to ask is whether there are areas of the universe more suitable for the birth of life. As we will see in Chap. 7, we now know that there are many planets in the Universe. In the next chapters, we discuss how they were discovered and their characteristics. Their position within galaxies is one of the factors contributing to their emergence. In 1925, Edwin Hubble classified galaxies according to their morphology. There are *elliptical galaxies*; in the shape of an ellipsoid, they are, on average, the largest and most massive. They are characterized by very old stars, are up to 12 billion years old, and have a red, orange color. The density is greatest near the center. *Spiral galaxies*, like ours, are characterized by an old and elliptical central bulb from which the spiral arms originate. The latter abound in clouds of dust and gas that give rise to new stars. Young (10 million years old) and blue stars predominate. Then, there are irregular galaxies that do not have a defined structure. Star formation is abundant. S0 galaxies, characterized by a bulge and a disk are more abundant than ellipticals in low-luminosity population, and are also more abundant than Spirals in galaxy clusters. Which of these galaxies are most suitable for life? Several parameters must be considered. First, for life to originate on a planet within a galaxy, it must be in a "comfort zone", an area with suitable conditions. These areas are called *habitable zones* or *Goldilocks zones*.

The name derives from the story Goldilocks in which a little girl is lost in the woods and arrives in an empty hut inhabited by a couple of bears with their little one. Everything in the hut is repeated in three different variations. There are three bowls with food, one very cold, the other very hot and the third at the right temperature. For the little girl, there are always two extreme choices, while only one is the right one for her. Like fairy tales, habitable zones are the right zones within which it is theoretically possible for a planet to maintain liquid water on its surface. They are also areas in which the "aggressions" of the galactic environment are minimal, allowing life to flourish. Giant black holes are found in the central area of galaxies. The X-ray and gamma radiation produced by the disk of material around the black hole are harmful to life. Supernovae are common in the galactic center. The initial flash of the explosion can destroy the ozone layer of nearby planets (up to tens of light years away), exposing life to cosmic rays and ultraviolet radiation. Supernovae are also frequent in spiral arms, areas of star formation and areas of high stellar density. In the past the passage of the Sun in the spiral arms was often correlated with mass extinctions, not only because of the presence of supernovae but also because, by passing inside the arms and the galactic disk, the comets found in the *Oort cloud* can be destabilized and head toward the center of the solar system. This is now an abandoned hypothesis. In addition to supernovae, *hypernovae* are even more dangerous, similar to supernovae, but with 100 times greater energy release. As we move away from the galactic center, the risk to life decreases, but at the same time, the quantity of heavy elements, indicated in astronomy with the term *metallicity*, decreases. A greater metallicity corresponds to a greater frequency of planetary formation, but exceeding a certain limit produces excess gaseous planets, which are generally not suitable for life. In the case of our galaxy, the habitable zone is a ring located between 15,000 and 38,000 light years (Fig. 6.2).

Below 15,000 years of light, the black hole and supernovae prevent the birth of life, and above 38,000 years of light, not enough rocky planets such as ours form owing to low metallicity. In 2015, the *Digital Space Exploration project* analyzed 150,000 nearby galaxies, creating a "cosmobiological model" for the habitability of the Universe. Pratika Dayal, an Indian astrophysicist, and her collaborators have shown that the galaxies with the highest metallicity are the largest and are usually elliptical. Furthermore, they can host 10,000 times more habitable planets than spiral galaxies. The concept of a galaxy's habitable zone can be extended to stars. It is therefore a ring around the star in which water is in a liquid state on the surface of a planet. Looking at the sky, you immediately see that the stars are not all the same; for example, you can see that they have different colors. Luminosity is related to temperature, which is in turn related to mass. The colors of a star are the result of the combination of emissions of different wavelengths. The hottest stars appear blue because they emit most of their energy in the blue part of the spectrum; fewer hot stars instead emit mainly in the red part of the spectrum. What is a spectrum? As Newton had shown many years earlier, if white light is passed through a prism, it breaks down into the colors of the rainbow. If we now consider a light source, pass it through a container with cold gas, and then pass the light through a prism, dark lines will be observed on the spectrum corresponding to the frequencies absorbed by the gas, forming an *absorption*

Fig. 6.2 Habitable zone of our galaxy. (Credits: NASA/JPL-Caltech/Lizbeth B. De La Torre)

spectrum. Each gas absorbs at precise frequencies, and the lines produced are similar to the fingerprints of a human being and are unique. They can be used to study which gas has absorbed the light. If we now consider a heated gas and pass the light through a prism, we will again observe lines with the same frequencies as the absorption spectrum. These lines constitute the *emission spectrum*. Stars are classified with the Harvard spectral classification, defined between the nineteenth and twentieth centuries. The latter classification is as follows: OBAFGKM RNS. In the classification, the stars range from the hottest and most massive to the least massive and red. For example, Orion is a blue star; it is class O and has a surface temperature of 30,000 degrees. The label is class B, is white–blue in color and has a temperature of 12,000 degrees. The brightest star in the sky, Sirius, is class A and has a white color and a temperature of 9900 degrees. Our sun is class G2 (yes, there are mixed classes, and each type is divided into subtypes from 0 to 9); it is yellow in color and has a temperature of 5700 degrees. Antares is an M-type star that is red in color and has a temperature of 2200 degrees. Which stars are best suited to life? The stars O, B, A up to F3 are too energetic and not long-lived; they are not suitable for supporting life. From F4 onward, together with the stars G and K, they seem the most suitable to support life owing to their stability, years of life, and duration of the habitable zone. G stars, like our sun, are conducive to life, but approximately half of them formed before there were enough heavier elements to form Earth-like planets. A quarter of them are younger than the Sun and therefore have had less time for evolution in their planetary systems to reach the levels of those on Earth. Around the stars there are zones where life is possible, called stellar habitable zones (Fig. 6.3).

F and K stars emit less ultraviolet radiation. In particular, K stars (orange dwarfs, 5% of all stars) are considered the most suitable for life. These stars are related to the *superhabitable planets*. Such superhabitable worlds are likely to be larger, hotter, and older than Earth. Since these stars are less luminous than the Sun is, their habitable zone is closer to the star. Erik Zakckrisson estimates that in the observable Universe, there are 7×10^{20} *Earths* (with masses between 0.5 and 5 times that of the Earth and radii between 0.8 and 1.5 the Earth's radius) and *super-Earths* (with masses in the range of 5–10 Earth masses and radii in the range of 1.5–2.5 Earth radii). Among these planets, 98% revolve around M-type stars, and only 20 billion revolve around those similar to the

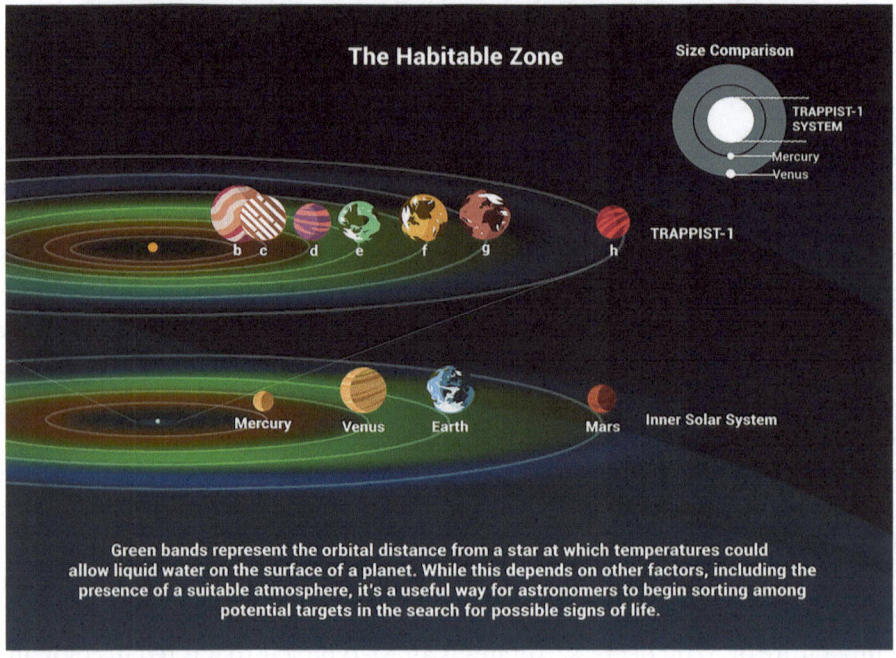

Fig. 6.3   Stellar habitable zone. (Credits: NASA)

sun. Different estimates come from Geoffrey Marcey, according to whom there would be 40 billion stars similar to our Sun and given that 22% of such stars have habitable planets, there would be 8.8 billion habitable Earth-like planets. Instead, according to a 2020 study, in our galaxy, there should be 6 billion rocky planets with dimensions similar to those of Earth that revolve around stars similar to those of the Sun. Ultimately, even considering the estimates with lower values, in our galaxy, there should be several billion Earth-type planets rotating around stars similar to the Sun.

If we then move on to M-type stars, these are the most abundant and have a very long life, meaning that there is more time for the appearance and evolution of life. A problem with these stars is that, in their first billion years, they are very active and are characterized by enormous emissions of radiation, including ultraviolet radiation. Another problem is that habitable planets would be located close to the star, and the gravitational interaction between the planet and the star would lead the planets to always show the same face. Therefore, one side would be heated continuously, and the other would not and would be very cold. On these planets, there would be only a narrow region where life would be possible. The smallest stars have a mass equal to 8% of the mass of the Sun. Smaller objects cannot trigger hydrogen fusion. Stars of this type are called *brown dwarfs* and have no habitable zones. During approximately 5 billion years of evolution, the Sun will become increasingly larger until it reaches Earth's orbit and will become a *red giant*. R-, N-, and S-type stars are carbon-rich stars and are red giants. Just as the Sun will not be able to support life in 5 billion years, neither will these stars. Beyond single stars such as the Sun, approximately 60% have companions. Is life possible in double or triple systems? This depends on how the planets are located relative to the stars. If the two stars are close and the planet is distant and external to them, the two stars behave as if they were one, and the situation is similar to that of single stars. The other possibility is that the two stars are distant and that the planet revolves around one of them. If the distance between the stars is greater than 5 times greater than that between the planet and the star around which it orbits, the situation is stable again. If, however, the stars are not far enough away from the planet, a stable situation is usually not achieved, and the planet can be expelled from the system. However, it would seem that most planets in systems with two stars belong to the first two cases; therefore, there may be a habitable zone. For a planet to host life, is it enough to be in the habitable zone? The answer to this question is a firm no. An example is Mars, which is in the habitable zone but does not appear to host life. There are many other factors that cause a planet in a habitable zone to host life. Eccentricity is an important factor. If it is too large and the planet moves very far from the star and at its maximum distance, temperatures could become very low. It seems that the rotation period also has a certain importance because temperatures should not change much between day and night. Another important factor is mass. Very large planets, which are usually gaseous, cannot support life because they lack a solid surface. Small planets such as "mini-earths", with a mass less than half that of the Earth, are unable to retain the atmosphere containing the volatile molecules typical of biogenic elements and are unable to have a hydrosphere or seas on the surface. Instead, the atmosphere will certainly be present on *Earth-like objects*, in *super-Earths* (mass between 1.9 and 10 Earth masses) and *mega-Earths* (mass greater than 10 Earth masses). Among the super-Earths, there could be so-called ocean worlds where water covers the entire planet. Another fundamental aspect is the nature of the atmosphere, which is the key to triggering the greenhouse effect, which warms the planet. In the case of the Earth, the greenhouse effect is due to the presence of water and carbon dioxide, which increase the average temperature by approximately 40 °C. Sara Seager claimed that another gas of considerable importance is molecular hydrogen, which is very effective, so much so that, according to Seager, planets that manage to maintain it can extend the limit of their star's habitable zone by up to 10 astronomical units. According to Seager, water can extend the habitable zone toward the star so much so that planets can exist up to 0.5 astronomical units from their star. Another very important factor is the magnetic field, which seems to be necessary for life. The presence or absence of a magnetic field on a planet also depends on its mass. The presence of high-mass planets such as Jupiter in the solar system could have an impact on life. Studies have shown that after the formation of the Earth, Jupiter hurled trillions of comets outward from the solar system, protecting the Earth. These conclusions, however, have been overturned by other scholars, since the existence of Jupiter did not allow the aggregation of materials between Mars and Jupiter to form another planet, and the asteroid belt was formed, which is dangerous for the Earth. Furthermore, comets and asteroids contain water and organic compounds of great probiotic interest. Another aspect that is often discussed is the role of satellites. According to some studies, the Moon is a type of stabilizer of the oscillations of the Earth's axis, which could reach 85° and we know that variations of 3° in the rotation axis can explain glaciations. The paleontologist Peter Ward and astronomer Donald Brownlee asserted that planets, planetary systems and galactic regions are suitable for complex life, as are the Earth, the solar system and our region of Milky Way, which are extremely rare, in accordance with the *rare Earth hypothesis*. It seems that moons as large as ours are not common, and this would confirm their hypothesis. More recent models by Jack Lissauer show that the stabilizing effect of the Moon has been overestimated and that without it, the axis wobble would not exceed 20°. We have also asked ourselves whether living outside habitable zones is possible. As previously mentioned, planets that retain molecular oxygen could contain liquid water even outside the habitable zones of their stars. Life could be present in some satellites of Jupiter and on Enceladus, a satellite of Saturn, owing to the presence of underground oceans and to processes such as those that take place in the areas of black fumaroles underwater. There is also the possibility that there is life that uses solvents other than water, as described in Chap. 9, as in the case of Titan, on which liquid methane flows. In this case, we discuss the *methane habitability zone*, which is very distant from the classical habitability zone located near the star. Therefore, life could not only have arisen in the habitable zones of stars but also outside them. If this happened, it would be proof of his "stubbornness".

7

The New Worlds

It is known that there is an infinite number of worlds because there is an infinite space capable of hosting them. However, not all of them are inhabited.
Douglas Adams

On 17 February 1600 in Piazza Campo de' Fiori in Rome, a philosopher, theologian and Dominican monk was burned alive on charges of heresy by the Holy Inquisition. It was Giordano Bruno, a Copernican. He was declared heretic because he was denying the eternal damnation, the Trinity, the divinity of Christ, the virginity of Mary, and transubstantiation. He was also a convinced believer in the fact that the Universe was infinite, that the stars were suns like ours around which planets revolved like in our solar system. These and many other ideas considered heretical were written in several books, such as *On the infinite, universe and worlds; The Ash Wednesday Supper; Cause, Principle, and Unity*, etc.

In his writings, we read

There is a single general space, a single vast immensity which we may freely call void: in it are unnumerable globes like this on which we live and grow, this space we declare to be infinite, since neither reason, convenience, sense-perception nor nature assign to it a limit

and again

I think of an infinite universe…. In fact, I consider something unworthy of the infinite divine power that, being able to create in addition to this world another and others, still infinite, had produced only one, finished. Therefore, I spoke of infinite Earth-like particular worlds

and again

However, we know that there is an infinite field, a space that embraces and interpenetrates the whole. In it is an infinity of bodies similar to our own. No one of these more than another is in the center of the universe, for the universe is infinite and therefore without center or limit, though these appertain to each of the worlds within the universe in the way I have explained on other occasions, especially when we demonstrated that there are certain determined definite centers, namely, the suns, fiery bodies around which revolve all planets, earths and waters, even as we see the seven wandering planets take their course around our sun. Thus, there is not merely one world, one earth, one sun, but as many worlds as we see bright lights around us …. In which other inhabitants move, live, vegetate and put into effect the acts of their vicissitudes.

The monk from Nola clearly managed, on the basis of the rudimentary knowledge of matter available in his time, to describe a physical theory of the Universe, which is very similar to what we know today. The description of the infinite Universe without center and margin is the basis of modern cosmology, and in his writings, he talks about other worlds and extraterrestrials living their lives on those worlds as we live on ours. Other thinkers, such as Bernard le Bovier de Fontenelle or Christian Huygens, in the same century, but after Giordano Bruno, continued to speak of a plurality of worlds. Strangely, as time progressed, these ideas were no longer accepted. For James Jeans or Arthur Eddington, the possibility of the existence of other planets was excluded. Giordano Bruno's ideas were far ahead and revolutionary for his time. Owing to the idea that other worlds exist in addition to ours and the verification that the idea was correct, observational proof is missing, and for this reason, we need to reach the 1990s. This is because it is almost impossible to see a planet rotating around the parent star given the enormous glow generated by the star. The light from the Sun, for example, is a billion times more intense than the light reflected from Jupiter. Even deducing its existence indirectly, as has been done, is not simple. Today, we know that there are several ways to do this, which we will now describe.

7.1 Hunt for Exoplanets

The first method is related to a very simple idea. We usually say that a planet revolves around its star. This is not exactly correct. The star and planet rotate around their common center of mass. Since a star is much more massive than a planet is, the center of mass will be inside it or near its surface. This is the case for the Sun and the other planets of the solar system. A few years ago, the Stephen Taylor group of NASA's *Jet Propulsion Laboratory* determined the position of the center of mass of the solar system with an accuracy of 100 m, and it was shown that it is located just outside the Sun. This means that the Sun will also move around the center of mass and oscillate around it. For example, Jupiter produces a displacement of approximately ten meters per second on the Sun in 12 years, whereas the Earth produces a displacement of 0.1 m/s per year. Determining such movements is absolutely not easy. The question that can be asked is how this shift can be determined. One possible way is to study the change in sunlight. How? *The Doppler effect was used.* A typical example to illustrate the latter is that of an ambulance. When it approaches, the siren's tone is higher, whereas when it moves away, it is lower. This effect works not only with sound waves but also with luminous waves and light. If a light source moves toward us, it undergoes an increase in frequency and appears bluer. When it moves away, the frequency decreases, and it seems redder to us. There is an instrument that allows one to establish how much the color of the star varies: the *spectrograph*. A spectrograph is an instrument that breaks light down into its basic components, wavelengths, and *spectrum*. As already mentioned, owing to the Doppler effect, if the source moves toward us, the lines in the spectrum will shift toward blue; if it moves away, the lines will shift toward red. By using it and measuring the color variations, the forward-backward motion of the star can be determined. In this way, the *radial velocity* is measured, namely, the speed of the star along the observer's field of view facing Earth. Herman Carl Vogel managed to measure this speed toward the end of the nineteenth century. In 1899, Vogel applied the idea to the star Spica and showed that it was a binary system. Although the companion star was too faint to be observed, spectroscopic analysis revealed that a companion was present. By the 1950s, trajectories of over 15,000 stars had been cataloged. The Russian-born American astronomer Otto Struve proposed in 1952 to use the Doppler effect to search for extrasolar planets.

Unfortunately, the technology of the time produced measurements of radial velocity with errors of 1000 m/s or more, making them useless for detecting orbiting planets. Geometry imposes another limit. The Doppler effect method is most effective when the planet containing the orbit appears in profile to the observer. At the other extreme, if there was an orbit perpendicular to our field of view, it could not be detected because the planet would not pull the star in our direction. Furthermore, a series of corrections must be made since the star is not a solid and homogeneous body. The presence of sunspots, for example, can hide the influence of a planet. Despite these problems, Struve proposed that the method could work for detecting planets. Struve died on April 6, 1963. In the same year, Peter van den Kamp believed that he had observed such an oscillation in *Barnard's Star*, discovered by the astronomer Edward Emerson Barnard in 1916. Located in the constellation Ophiuchus, Barnard's star is so dim in visible light that it cannot be observed with the naked eye. It has a peculiar characteristic: it has greater proper motion than any other known star does, and for this reason, it is also called *Barnard's Arrow Star*. van den Kamp had begun his activity as a planet hunter as early as 1938 together with his colleagues at the Sproul Observatory at Swarthmore College and continued throughout his career. To avoid individual errors, photographic plates acquired at the Sproul Observatory were shown to an average of ten people each. In 1963, the astronomer declared that around the star, there was a planet similar to Jupiter, with a mass 1.6 times the mass of Jupiter and at a distance of 4.4 astronomical units from Barnard's Stars. In 1969, the result was confirmed, and in the same year, he published another article in which he argued that there were 2 planets, one with a mass of 1.1 Jupiter masses and the other with a mass equal to 0.8 Jupiter masses. Before van den Kamp, there were controversies about the binary star 70 Ophiuchi. The discovery of van de Kamp received wide credit in the astronomical community between 1963 and 1973. In 1973, George Gatewood and Heinrich Eichhorn used two different measurement techniques on 241 photographic plates acquired at the Allegheny and Van Vleck observatories and disproved Van de Kamp's discovery. Four months after the work of Gatewood and Eichhorn, John L. Hershey published an article relating the change in the position of Barnard's Star to the modifications and adjustments that had affected the lenses of the Sproul Observatory telescope in 1949 and 1957. van den Kamp had arrived at an incorrect result due to an instrumental error, but he never acknowledged that he had made errors and continued to believe in the validity of his discovery, which he reiterated in subsequent articles, the last of which was in 1982 and in an interview from 1985. The instrumental errors in the photographic plates of the Sproul observatory led van de Kamp and colleagues to announce the discovery of planets around other stars, such as Lalande 21185, 61 Cygni and others, discoveries that were later refuted. However, the story surrounding Barnard's Star does not end here. Recently, in 2018, an article in Nature reported the existence of a planet that is at least 3.2 times more massive than Earth, a result of the *Red Dots* and *CARMENES* projects, which were disproved in 2021. The search for extrasolar planets had actually already begun in 1855, when Captain WS Jacob measured anomalies in the binary star 70 Ophiuchi in orbit, so much so that he considered it "highly probable" that these anomalies were due to the presence of a planet. Between 1896 and 1989, Thomas J.J. Seehe argued that the anomalies were due to the presence of a dark companion with an orbital period of 36 years connected to one of the two stars of the binary system. This thesis was opposed by Forest Ray Molton, who, in 1899, published his own analyses according to which a three-body system with the orbital parameters described by See would be highly unstable.

An important discovery was that of the circumstellar disk around the star β *Pictoris*, which represents the region in which the formation of new planets is underway or the residues of this process.

The history of errors in the search for extrasolar planets continued with several other cases. The first was the discovery in 1989 by David Latham's group of an object having a mass, in their estimate, equal to 11 times that of Jupiter, which revolved around the star HD 114762. There was doubt that it was a brown dwarf; however, the discovery was confirmed only in 2012, and in 2022, it was estimated that the mass is 147 times greater than that of Jupiter; therefore, it is not a planet but a *brown dwarf*. The second was the discovery of a planet by the English group of Jodrell Bank led by Andrew Lyne. In July 1991, the quoted group announced the discovery of the first extrasolar planer around a "dead star", a pulsar named PSR 1822--210. I will speak of these stars in a while. After the discovery, it was clear that some people had forgotten to update the position of the PSR from 1822--10. When the position was corrected, it was clear that the planet was not present. In the meeting of the American Astronomical Society of January 1992, Lyne had to admit the mistake. In the same meeting, a young astronomer, Alex Wolszczan, announced the first planets discovered around another pulsar, PSR B1257+12. This time, the discovery was real but did not cause much sensation and was not celebrated by the mass media because the star, as previously reported, was not a "normal" star but a *pulsar*, namely, a *neutron star*. What is it? A star is formed by the gravitational collapse of dense, cold clouds of molecular gas. As the collapse increases, the cloud shrinks, and its central temperature increases until it reaches values such as triggering nuclear fusion, which involves the fusion of four hydrogen nuclei to form a helium atom. Energy is released in the process. The star is in equilibrium under the action of gravity, which tends to cause it to collapse, and under gas and radiation pressures, which tends to cause it to expand. This equilibrium continues as long as there are fusion reactions. In the case of stars with a mass lower than 8 solar masses, when a good part of the hydrogen in the core is consumed, the radiation pressure capable of balancing gravity decreases, and the central part of the star contracts, triggering hydrogen in the shell around the center. Owing to the higher temperatures, the rate of nuclear reactions is greater, causing the star to brighten by a factor of 100–1000. The increase in the density of the core and its temperature translates into an expansion of the surface layers of the star. Because the energy produced is released over a larger surface area and because some of it is dissipated in the expansion, this results in a lower surface temperature of the star. The star becomes a *red giant*. The life of a star is prolonged by the ignition of helium in the core. When the helium also runs out in the core, there will be a new expansion of the outermost layers and a contraction of the innermost layers, allowing the helium to fuse into a shell around the center. Stars with a mass lower than 8 solar masses, however, do not have a mass large enough to reach the temperature and pressure necessary to fuse carbon and the carbon, no longer supported by radiation pressure, will collapse under its weight and expel most of its mass, forming a *planetary nebula* (Fig. 7.1).

Only the core will remain to form a star known as a *white dwarf*, with a mass similar to the Sun, a size a thousand times smaller and a density a million times higher than that of the Sun. Today, many white dwarfs are known. The first was discovered by Friedrich Wilhelm Bessel, a great German mathematician and astronomer. Sirius is the brightest star in the sky. Its motion does not follow a straight line but rather twists in a serpentine motion. This betrays, in accordance with Newton's laws, the presence of a companion not easily visible, at least with the technology of Bessel's time. The companion, called Sirius B, was discovered twenty years later. It was expected that the star, being not very bright, would be red, but instead, in 1915, it was seen that it was white, with a mass similar to that of the Sun. If the mass of the star is greater than 8 solar masses, the temperature and pressure at the center are so high that they fuse elements *that are* heavier than carbon. Carbon fusion reactions produce several products, mainly sodium, magnesium, oxygen and neon. *Like in* the case of helium, depending on the mass, the fusion of carbon can occur with a flash or not. This effect, the expansion of the star and the increase in brightness *cause* the radiation pressure to produce stellar winds, and the star gradually loses mass. When the temperature reaches 1.8 billion degrees, oxygen fusion reactions are produced, which *produce* elements important for life, such as sulfur and phosphorus. Photodisintegration processes of magnesium

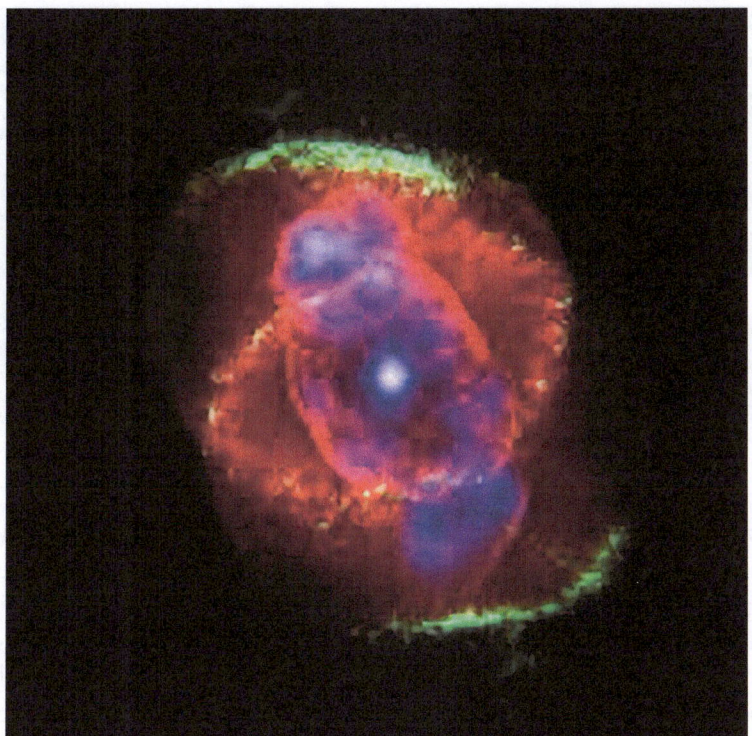

Fig. 7.1 Planetary nebula Cat's eye. (Credits: NASA)

begin, and those of silicon begin at 3.4 billion degrees, leading to the formation of more stable nuclei, such as iron and its isotopes, including ^{56}Fe (Iron-56), which is extremely stable. If you want iron to form a heavier element, you need to supply more energy than is released by fusion. When it reaches this point, the nuclear energy source at the center of the star is consumed. The star has an onion structure: in the center, there is iron, and going outward, there is silicon, oxygen, carbon, helium and hydrogen. When it reaches this point, the nuclear energy source at the center of the star is consumed. Gravity is not hindered by anything and can cause the star to collapse. Owing to their high pressure, protons capture electrons, forming neutrons. These processes (rupture of nuclei, formation of neutrons and emission of neutrinos) consume energy and therefore favor the collapse of the nucleus, forming a *neutron star* approximately 10 km in size (Fig. 7.2).

In this case, the star will produce a *large* explosion with an intensity that can surpass *that of* an entire galaxy. This state is referred to as *the supernova state*. In the event *that* a large amount of material is hurled into space, *a*

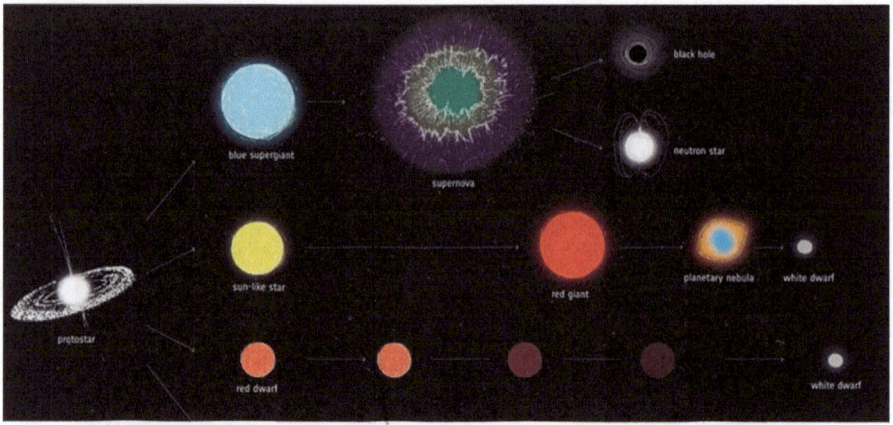

Fig. 7.2 Principal phases of stellar evolution. (Credits: ESA)

supernova remnant is formed. Today, several supernova remnants are known, and one of the most famous is the famous *Crab Nebula.* This is the remnant of a supernova that exploded in 1054, as observed by Chinese astronomers, and was so bright that it was observed for several months. The initial rotation of the star, which is in collapse, is amplified because of the mechanism described when a skater brings his arms closer to the body. In the case of the neutron star inside the Crab Nebula one revolution occurs every 0.03 s. The rotation together with the magnetic field produces an emission of pulsed radiation, such as that of a lighthouse, visible from very far away, and the emission is so regular that it marks time with a precision greater than that of an atomic clock. Pulsars were discovered in 1967 by Jocelyn Bell in Cambridge. There are pulsars that rotate much faster than the crab and millisecond pulsars. These are very old pulsars that no longer emit pulses because over time, they have lost energy and slowed down. Somehow, they manage to attract material from nearby stars and be reaccelerated. In 1983, Alexander Wolszczan worked at the Arecibo radio telescope in Puerto Rico. Since the telescope broke and was repaired after two years, it could not be oriented toward the galactic plane. Wolszczan could then use the radio telescope at will and observe unusual directions in space in search of pulsars. He discovered the pulsar PSR B1257+12 in the constellation Virgo. With the help of Dale Frail, who worked at the *VLA* (Very Large Array), which is a telescope made up of 27 antennas in New Mexico. Combining the signals from all the telescopes in December 1991, Wolszczan and Frail reported disturbances in the signal from the pulsar and reported that these disturbances were due to two planets with masses no less than 3.4 and 2.8 times greater than those of Earth and orbiting, respectively, at 0.36 and 0.47 astronomical units around the pulsar PSR B 1257+12. In 1994, a third planet was identified with a mass equal to twice the Moon and orbiting at 0.19 astronomical units. Now, a problem arises. How was it possible that after the large supernova explosion, the remaining pulsar had planets around it? The only possibility was that they had formed, in the same way that planets formed, from material left after the explosion. However, the enormous amount of radiation from the pulsar made it clear that the three planets certainly could not host life. This discovery highlighted the fact that if there were planets even around a pulsar, there must be planets around ordinary stars. In fact, on October 5, 1995, Michel Mayor and Didier Queloz of the Geneva Observatory announced the discovery of an extrasolar planet with a mass similar to that of Jupiter rotating around the star 51 Pegasi.

7.2   51 Pegasi

The observations of the two astronomers were carried out at the Observatory of Haute Provence, which is located on a hill and is not equipped with a very powerful telescope with a 1.93 m aperture. The two astronomers bought observations at the quoted observatory because the telescope was available for long periods of time. The strong part of the two astronomers' equipment was the *spectrograph* Elodie, developed at the Marseille Observatory. The device was connected to a typical computer. The two astronomers used Struve's idea, which we have already discussed. By breaking down the light with the spectrograph, they observed the movement of the spectral lines. If the source moves toward us, the lines shift toward blue; if it moves away, the lines shift toward red because of the Doppler effect. We have seen that the search for extrasolar planets had already begun in 1855, but at that time and in the following decades, the technology was not capable of determining the movement of the spectral lines. In the 1980s, technology improved greatly, and many astronomers dedicated themselves to searching for planets. The main groups of interest in this research were the groups of Bruce Campbell and Gordon Walker, from British Columbia; Geoffrey Marcy and Paul Butler, San Francisco; Artie Hatzes and William Cochran, Texas; and later, the Swiss groups of Mayor and Queloz. The two Swisss started in April 1994, several years after the other groups did, but luck was on their side. The search for an exoplanet is conceptually very simple. First, we need to observe the shifts of the spectral lines, and then, we need to perform calculations to determine the parameters of the planet. One may wonder why the Swiss arrived before their competitors, who left early. The answer is that they were able to do the two jobs indicated very well. Marcy and Butler had more precise instruments; they were able to measure stellar motions of 3 m per second, whereas Swiss was only 13 m per second. However, Americans slowed down in the second phase, the calculation phase, because they used a calculation algorithm inferior to that of the Swiss. The two Americans decided to make many observations and record the results, and at the end, they would make the necessary calculations. In September 1994, the Swiss began to study a Sun-like star 50 years away from us, the star 51 Pegasi, which, as its name suggests, was located in the constellation of Pegasus. In December 1994, the Swiss spent a week with 51 Pegasi and noticed some strange things. It appeared to be moving in a circular orbit at a speed of 60 m/s, a very large speed never previously observed by any group. They thought there were problems with the spectrograph. They tested it on other stars, and it seemed to work wonders. Therefore, the phenomenon observed was typical of 51 Pegasi. In January, they resumed observations of the star and established that an object with a mass approximately 60% greater than that of Jupiter must be moving around it, and the most absurd thing was that the object was moving in an orbit smaller than that of Mercury in our solar system. This contradicts the theory of planetary formation mentioned in Chap. 4. According to this theory, there is a line, the *ice line*, which indicates the distance within which materials can condense to form rocky planets. Beyond this line, giant gas planets have formed. In the solar system, this line is located 2.7 astronomical units from the Sun. Obviously, the position of this line depends on the characteristics of the star, but 51 Pegasi is similar to the Sun; therefore, it was absurd to find a planet such as Jupiter at distances from the star lower than those of Mercury in our study. In the following years, this oddity was explained by the phenomenon of *planetary migration*, according to which a planet is born in a position on the disk and can move owing to the interaction of the planet with the disk. This happens for the most massive planets. Even in our solar system, there were probably migrations of Jupiter and Saturn, but of small magnitude. By March 1995, Mayor and Queloz were nearing the end of their search. On July 4, 1995, they made one last observation late at night, four months after the last observations, which confirmed their results. Approximately 51 Pegasi revolved a planet with a mass of 60% of the mass of Jupiter, which moved in a very narrow orbit of only 4.23 days approximately 51 Pegasi. Because they were almost sure of the existence of the planet, they had brought their families and celebrated with them the great event. The results are summarized in an article that was presented at a conference in October in Florence. Marcy and Butler verified the Swiss result a few days after the congress. Reviewing data that had been taken for years but not checked, Marcy and Butler realized that they had discovered exoplanets before the Swiss. They were mocked by their choice to wait and read the data later. In the following years, several gas giants orbiting near their stars were discovered, with planets similar to those of 51 Pegasi. These planets are referred to as *Hot Jupiters*. The quantitative supremacy of these planets over the others was caused by the selection effect of the radial velocity method. With improvements in instruments and the use of other techniques, plans very different from these have been revealed. In 1999, the first multiple planetary system was discovered around the star *Upsilon Andromedae,* and in the same year, the transit of a planet in front of its parent star, HD 209458 b, was observed. As already mentioned, Mayor and Queloz's discovery was made with an instrument that could appreciate variations of 13 m/s. With the new *ELODIE spectrograph,* it decreased to 7 m/s, and with *SOPHIE, it increased* to 3 m/s. A spectrograph, *CORALIE,* was also installed in the Southern Hemisphere at the La Silla Observatory in Chile. CORALIE was then replaced with *HARPS,* which reached the limit of 1 m/s and, over time, reached 0.5 and 0.2 m/s in short-distance observations, a thousand times less than Otto Struve thought possible in the 1950s. In 2009, an analysis of the star Gliese 581 revealed the existence of a planet with a mass 1.9 times the mass of Earth. In 2014, again with the radial velocity technique, a planet was discovered, Kepler-138 d, with a mass equal to that of the Earth but which rotated very close to its star. Moreover, the *HARPS-N* system was installed at the *Galileo National Telescope,* which is located in the Canary Islands. *ESPRESSO* was installed in the *VLT* (equipped with 4 telescopes with 8.2 m aperture mirrors) in the Atacama desert in Chile, which can reach a precision of 10 cm/s and allows it to detect planets with masses and orbits similar to those of the Earth. When the construction of the *E-ELT telescope* (European Extremely Large Telescope) with an aperture mirror of 40 m is finished, it will be possible to achieve precision of centimeters per second through the use of the *CODEX* spectrograph.

7.3 Occultations and Other Planet Traps

The radial velocity method, as already mentioned, is more sensitive to large planets close to the central star. There are other methods without this limitation. If the plane of the orbit of an extrasolar system is edgewise with respect to us, as the planet moves, it will occult part of the star, causing its brightness to decrease. This method is the *method of transit* (Fig. 7.3) or *occultation.*

The parameters that affect the observation of planetary transit are the size of the planet and the period of revolution. A single observation of a transit gives indications of the existence of a planet but does not provide reliable

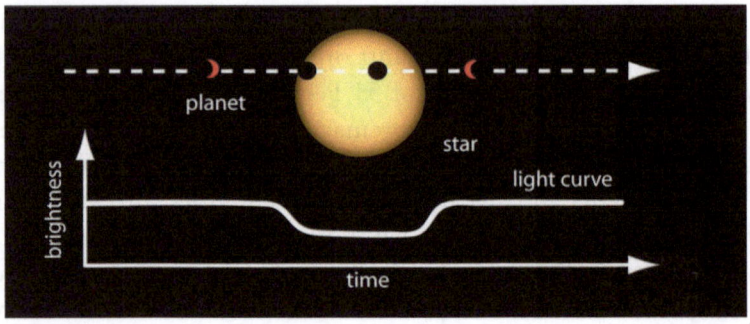

Fig. 7.3   Method of transit. (Credits: NASA)

information on the rotation period. It is therefore necessary to proceed with the observation of two periods. However, this can take a long time. In the case of Jupiter, who has an orbital period of 11.86 years, the observer would have to wait at least 36 years. In other words, with the transit method, it is easier to observe large planets close to the central star. The first planet discovered via this method was *HD 209458 b*, which caused its star to decrease in brightness by 1.5% in 1999. HD 209458 b has a radius 40% larger than that of Jupiter and a rotation period of half a week. The greatest drawback of the transit method is that it does not allow the mass of the planet to be determined unless it is combined, for example, with the *radial velocity technique*. This was done for HD 209458 b and was found to have a mass 0.7 times greater than that of Jupiter. The transit method, however, has two notable advantages: being able to measure the size of the planet and not requiring large telescopes to measure the decrease in the brightness of the star. A group from the Harvard–Smithsonian Center developed the *Mearth project*, with eight 40 cm aperture telescopes. In 2009, the planet *Gliese 1214b* was discovered. *SuperWASP* is a project similar to the American one, in which there are two observatories, such as the American one with 10 cm aperture telescopes. Obviously, in space, a place far from artificial lights, atmospheric dust and air refraction, the situation is even better. In 2006, the European and French Space Agency launched *Corot*, which discovered approximately thirty planets, including *Corot-7 b*, which has a mass equal to 1.7 Earth masses and ended its activity in 2012. In 2009, it was replaced by a NASA satellite, *Kepler*, with the mission to search for and confirm Earth-like planets orbiting stars other than the Sun. The mission's expected time was initially 3.5 years, but it was repeatedly extended until it officially ended in October 2018. During its period of activity, it observed 530,506 stars and detected 2662 planets, some of which had conditions potentially suitable for life. In 2018, NASA launched Kepler's replacement, the *TESS* (Transiting Exoplanet Survey Satellite) orbital telescope, which is considered Kepler's successor. While Kepler examined a limited portion of the celestial vault, approximately 0.28%, TESS examined it all, focusing on stars from thirty--one hundred times brighter than those observed by the predecessor. There is also a European Space Agency project intended for the study of exoplanets, *CHEOPS* (Characterizing Exoplanets Satellite), with the main scientific objective of studying the structure of exoplanets with radii that typically range from 1 to 6 times those of the Earth and with masses up to 20 times that of our planet, orbiting around bright stars. The European Space Agency is also preparing the *PLATO* (Planetary Transits and Oscillations of Stars) mission, a satellite equipped with 34 small 12 cm telescopes dedicated to the search for exoplanets around bright stars, with the main objective of identifying exoplanets similar to the Earth through the transit method and measuring the oscillations of stars around their orbit to determine their mass, radius and age with unprecedented accuracy. In addition to the two quoted methods, an exoplanet can be discovered via other techniques. One of the methods is closely related to that of radial velocities. As already mentioned, a star and a planet move around the center of mass, which causes the star to oscillate. The oscillation can be measured with the *Doppler effect* or directly with the *astrometry method*. This method was used, as already mentioned, for the first time by Bessel. Observing Sirius, he noticed a nonrectilinear motion that led him to think that the star had a companion that was revealed with direct observation in 1862. If we consider the Sun-Earth system, excluding the other planets, the center of mass of the system would be 450 km from the center of the Sun. The Sun and Earth would move around the center of mass, but since it is almost at the center of the Sun, the latter would move by an imperceptible amount. By replacing the Earth with Jupiter, the center of mass would be just outside the Sun, and the oscillatory effect that Jupiter causes on the Sun would be 1650 times greater. In the case of astrometry, the effect of the planet depends on the product of its mass and the distance from the star. Therefore, this technique provides optimal results in the case of planetary systems with nearby stars and with massive planets orbiting far from the main star and with long-period orbits. In 2013, the European Space Agency launched the *GAIA* orbital observatory. The observatory is obtaining astrometric data of more than a billion stars with two hundred times greater precision than its predecessor *Hipparcos*. According to some estimates, by the end of the mission, GAIA may have revealed hundreds of thousands of exoplanets. Another method is based on the predictions of Einstein's general relativity. Owing to the deformation of space-time masses, the light deviates, and the *gravitational lens effect* occurs. The effect was verified for the first time in an eclipse in 1919. In 1978, an Anglo-American group observed a pair of quasars that later turned out to be the double image of a single quasar due to the gravitational lensing effect. In 1985, the so-called *Einstein Cross* (object G2237+0305) was observed, i.e., a gravitational lens showing four images of the same quasar around a galaxy. In 1998, the collaboration of a team from Manchester and the Hubble Telescope revealed the first Einstein ring (B1938+666). This object is obtained from the annular deformation of light coming from a galaxy. Even in the case of stellar objects, the lensing effect can occur, or more precisely, the *gravitational microlensing effect* (Fig. 7.4) as proposed by Bohdan Paczynski in 1986. If from Earth you observe, for example, a star in the Magellanic Cloud and a body cuts the line between the observer and the star, an increase in starlight is observed.

In 1990, Paczynski started the optical gravitational lens (OGLE) experiment. From 1992 to 2009, more than 4000 events related to quasars, binary stars, etc., were detected. In 2004, the experiment was improved and allowed the detection of planets, the first being a planet 2.6 times more massive than Jupiter and 17,000 light years away. Other similar experiments, such as *MOA* (Microlensing Observations in Astrophysics) and *KMT-Net* (Korea Microlensing Telescope Network), have been conducted. The microlensing method allows the discovery of very distant exoplanets, such as the one discovered in 2015, which was 27,700 light years away. A space telescope has been designed, NASA's *WFIRST* (Wide

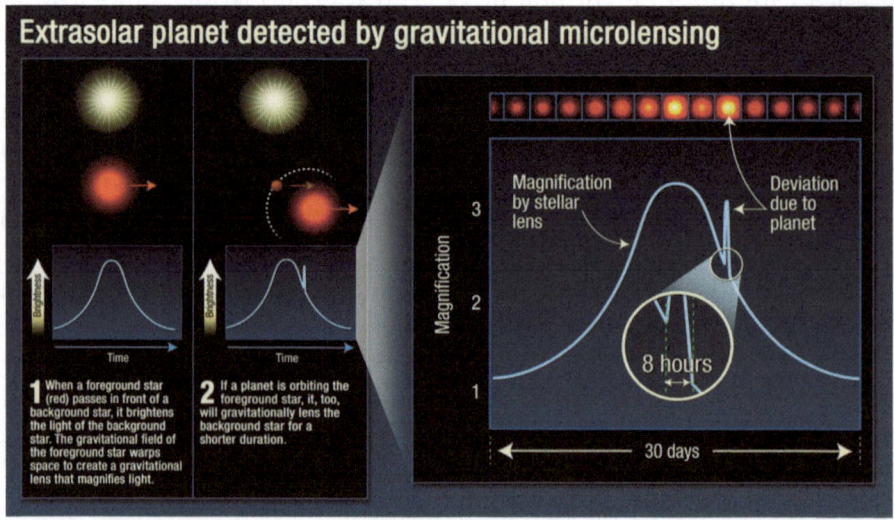

Fig. 7.4 Gravitational micro-lens technique. The light curve of a microlensing event without (left) and with (right) a planet orbiting the star. The peak locates the planet. (Credit: NASA, ESA, and K. Sahu (STScI))

Field Infrared Survey Telescope), which should be launched in 2025 and which will 'hunt' planets, even distant and small ones, with the gravitational lens method. Another method is *pulsar timing*. Already discussed when we talked about the discovery of the first exoplanets from Wolszczan. Finally, *direct revelation* remains. Surely if we could directly observe a planet, it would be a good thing, for example, because we could try to study its atmosphere trying to understand if it hosts life. Direct observation is very difficult, as can easily be understood, because the reflected light of a planet is substantially less intense than that of a star. The light from the Sun is ten billion times more intense than that from the Earth in the visible region, but if we consider the infrared region, it is only ten million times more intense. To improve things even further, one can try to block the star's light using an artificial blocking system, called a *coronagraph*. The first directly observed exoplanet, in 2004, was 2M1207, 170 years away from Earth, having a mass several times greater than that of Jupiter and rotating around a brown dwarf only twenty times brighter than the planet. This result was obtained by a French-American team that used an infrared observation instrument coupled to the VLT. In 2008, it was possible to observe an exoplanet with a mass 3 times that of Jupiter in the visible rotation around the star *Fomalhaut* with the Hubble Space Telescope. In 2009, the analysis of images dating back to 2003 revealed a planet orbiting *Beta Pictoris*. In 2012, a planet with a mass of 12.8 Jupiter was observed orbiting *Kappa Andromedae* with the *Subaru telescope*. In 2015, a gas giant was discovered orbiting *51 Eridani*, 96 light years from Earth. The number of planets directly observed is now close to one hundred, and in almost all cases, they are gas giants, with masses greater than Jupiter. Radio telescopes will also be used to search for extrasolar planets. The Chinese five hundred meter aperture spherical telescope (FAST) is currently the largest and most sensitive radio telescope in the world and is three times more sensitive than the Arecibo Observatory radio telescope. This instrument will be used to detect extrasolar planets, and when it is completed, the Square Kilometer Array (SKA) radio telescope will also contribute to the search.

We currently have several methods for detecting extrasolar planets that have proven to be very effective. As of March 27, 2024, the *Encyclopaedia of Exoplanetary Systems website* reported the discovery of 5652 exoplanets. This number, although significant, is very small compared with the estimates of exoplanets in our galaxy, which, as already mentioned, should be hundreds of billions. The discovered planets are of different types, from Hot Jupiters to Earth-like planets, which are the planets that interest us most in the search for life in the cosmos. What types of planets have we discovered, are there any that could be habitable? We discuss this in the next chapter.

8

Is There Life on Extrasolar Planets?

At the beginning of the 1990s, astrophysicists thought that, owing to the advancement of technology since the time of Otto Struve and owing to some luck, they would be able to identify exoplanets around nearby stars. In addition, as we have said, that was precisely what happened with the discoveries of Wolszczan and then those of Mayor and Queloz. Today, more than 5000 planets have been discovered, with distances between 4.2 and 27,700 light years. Some are less massive than the Moon, whereas others have masses close to that of a brown dwarf and radii in the range of 2000–200,000 km. The variety of exoplanets discovered is so vast that we cannot use the classification used for our solar system, and at the same time, they are not representative samples of those existing in our galaxy. Our interest is fundamentally in the habitability of these planets. Therefore, we will determine whether there are habitable planets in the groups into which they can be divided.

8.1 Jupiter-Type Planets

The first group of objects is that of Jovian-type planets, with masses similar to those of Jupiter and Saturn. A type of planet that does not exist in our solar system is the *hot Jupiters*, planets with Jupiter masses but very close to the star, whose existence, as we have already noted, can be explained by *planetary migration*. That is, they were born far from the star and then slowly "slipped" toward it. 51 Pegasi b, the first exoplanet discovered, is a hot Jupiter. These objects are obviously not interesting as far as life is concerned. They are gaseous, they do not have a surface, and furthermore, they have very high temperatures on the illuminated side and are very cold on the other side. Given the high temperatures, the fog existing on these planets is made up of silicates and molten iron. Then, there are the *cold Jupiters*, i.e., in a situation similar to our Jupiter, and then the *eccentric Jupiters*, their orbit is very eccentric, that is, it is very elongated. All three types of planets cannot be habitable, and furthermore, the hot Jupiters and eccentric Jupiters, with their movement in the system, do not allow the formation of terrestrial-type planets where life could form. Therefore, these planets have no possibility of giving birth to life.

8.2   Neptune-Like Planets

These planets have masses intermediate between those of Earth and Jupiter, with masses between 1 and 50 times greater than those of Earth. Additionally, in this case, there are *hot Neptunian planets* and *cold Neptunian planets*. *Gliese 436 b* is a hot Neptunian planet that rotates just 4 million kilometers from the star and has a mass 30% greater than that of Neptune. These planets are also harmful to life because they can migrate like Jovian-type planets and are also not suitable for hosting life.

8.3   Mini-Neptunes and Super-Earths

Mini-Neptunes are planets similar to Neptunes, with a mass 10 times greater than that of Earth, and therefore have a central rocky core and a thick mantle of gas. *SuperEarths* have masses between 2 and 10 Earth masses. However, they do not have the same structure, composition and habitability as the Earth does. The first extrasolar planets discovered are indeed super-Earths. They are the first two planets discovered by Wolszczan around the pulsar PSR B1257+12, i.e., *PSR B1257+12 c*, and *PSR B1257+12 d*, with masses of 4.1 and 3.8, respectively, of the mass of the Earth. These planets are obviously not habitable. The radiation coming from the pulsar is absolutely lethal. Gliese 876 c, d, and e are superEarths orbiting a red dwarf. *Gliese 876 c*, with a mass 5 times greater than that of the Earth, appears to have conditions similar to those of Venus, whereas *Gliese 876 d*, with a mass 7.7 times greater than that of the Earth, orbits within the habitable zone, corresponding to its external limit. *Gliese 876 e* has a mass of 1.9 Earth masses and orbits its star in 3.15 days at an average distance of 0.03 astronomical units. The planet is believed to experience at least 100 times more tidal heating than Jupiter's satellite *Io* experiences. There is a long list of super-Earths, 288, discovered by Kepler. It is possible that a super-Earth is more similar in structure to a red giant than to Earth. Obviously, the search for extrasolar planets aims to find objects that are similar to Earth not only in their mass and size but also in their ability to host life.

8.4 Exo-Earths

Planets that resemble Earth not only in mass and size but also in ability are so-called *exo-Earths*. In 2009, *Gliese 581 e* was discovered, an exoplanet with a mass two times less than that of the Earth, located 4 million kilometers away so that it could host some forms of life. In 2010, Steven Vogt, together with others, discovered *Gliese 581 g*, which he named *Zarmina*, after his wife. Being in the Gliese habitable zone and being the most Earth-like planet ever identified until then, it was believed to have the potential to host life. It is a shame that its existence was disproved in 2014. Kepler has discovered many planets with dimensions similar to those of Earth, with radii slightly larger, and in some cases, it has even identified planets smaller than Earth, such as those orbiting around *Kepler-42* or *Kepler-20 e*, the first discovered exoplanet smaller than Earth orbiting a solar-type star. According to a 2020 study by Michelle Kunimoto and Jaymie M. Matthews, there should be 6 billion Earth-sized rocky planets in our galaxy revolving around Sun-like stars.

8.5 Habitable Planets

So the estimates for the number of habitable planets are very positive, but have any been revealed? Yes, today, there are more than fifty. The turning point came with the launch of the Kepler mission in 2009. To have a greater chance of finding habitable planets, Kepler was aimed at an area of the sky far from the ecliptic, avoiding dust and disturbances from the asteroid belt and Kuiper. A field was chosen near the constellation, which is suitable for avoiding the central regions of the galaxy. Earth-based observations have also contributed, such as the *TRAPPIST* (Transiting Planets and Planetesimals Small Telescope-South) observatory, a 60 cm robotic telescope installed at the La Silla Observatory in 2010. Michael Gillon and colleagues used a telescope to observe red dwarf stars in 2015. 2MASS J23062928-0502285, which is now also known as *TRAPPIST-1* (Fig. 8.1) discovered three terrestrial planets with the outermost planet appearing within the small star habitable zone.

Fig. 8.1 TRAPPIST 1 system and solar system. (Credits: NASA)

In 2017, the *Spitzer Space Telescope* studied the star and discovered 4 other planets, some of which are located in the habitable zone. Describing all the planets in the habitable zone would be long and tedious, so we only consider a few, those that have a rocky composition, with masses lower than 6 Earth masses. The parameters used to establish similarity with the Earth and habitability are the PHI index and the ESI (Earth similarity index). The ESI is more relevant for exoplanets than the PHI, for which there is not much data on habitability. Most of the planets found in the *conservative habitable zone* (part of the habitable zone where conditions remain favorable for most of the life of the star) are planets that revolve around red dwarfs. A problem common to these planets is the fact that they rotate around the star in *synchronous rotation*, that is, always turning the same hemisphere toward the parent star. This causes one side to be very hot and the other to be cold. In this type of planet, life should be concentrated in the area of the *terminator* or *circle of illumination*, the fictitious line that delimits the illuminated part from the shadowed part. Red dwarfs also face the problem of being subject to violent flares, especially in the youthful phase, which can cause major problems for life. The highest ESI planet discovered thus far is *Teegarden b*, with an ESI of 0.95, revolving around the red dwarf star, Teegarden. Teegarden b rotates very close to the star in *synchronous rotation*. The Teegarden star is approximately 8 billion years old and should not have the problem of flares. The rotation around the star is *Teegarden c*, which has an ESI of 0.68 and should be similar to Mars, even if it has a greater mass. Another planet that is in the conservative habitable zone, with an ESI equal to 0.93 and with a mass at most twice that of the Earth, which rotates together with two other companions around the red dwarf TOI 700, is *TOI 700 d*. On a planet with an average temperature of 22 °C, liquid water should flow, but it presents the problem of planets rotating around red dwarfs. *Kepler-1649 c* has an ESI of 0.92, has a mass very similar to that of Earth, and perhaps its average temperature is also similar to Earth's temperature. The composition of the atmosphere is not known, and consequently, it is not known whether liquid water is present. The parent star is a red dwarf, with flares that could severely hinder the development of life on the planet. Seven planets rotate around the red dwarf 2MASS J23062928-0502285, also known as *Trappist-1*. The most interesting aspect of life is *TRAPPIST-1d*, which has an ESI of 0.91. It is smaller and less massive and dense than Earth and should have a temperature of approximately 17 °C. The density could indicate the presence of large quantities of liquid water in the form of oceans. TRAPPIST-1 is also rocky and has dimensions similar to those of Earth. According to one study, this would be the planet with the greatest probability of life in the TRAPPIST-1 system; however, the results contradict those of other studies. *LP 890-9 c*, which has an ESI of 0.89, also revolves around a red dwarf that is 7 billion years old and therefore should be stable. The planet will probably be observed by the James Webb Space Telescope to study its atmosphere. The closest habitable planet to us is *Proxima Centauri b,* which has an ESI of 0.87. Its mass between 1.17 and 3 Earth masses suggests that it may be a terrestrial planet. There is no certainty about habitability, but the fact that the parent star is a red dwarf implies two usual problems: synchronous rotation and strong flares. The super flares observed in 2017 and subsequent studies on Proxima b led us to believe that the planet is not the best candidate for searching for extraterrestrial life forms. *K2-18b* seemed to be a planet of considerable importance because, in 2023, observations with the James Webb telescope revealed the presence of a molecule produced only by life, *dimethyl sulfide*. Unfortunately, in May 2024, simulations revealed that the dimethyl sulfide signal overlaps with that of methane, and the situation is therefore unclear. In any case, the atmosphere contains methane and carbon dioxide, which together are strong indicators that favor life. Other habitable planets revolving around red dwarfs are *K2-72* and *Gliese 1002 b, Gliese 1061 d, Ross 128 b, Gliese 273 b*, etc. Planets have also been discovered around stars that do not have the problems of red dwarfs. *Kepler-452 b* with an ESI of 0.83 is the first planet with dimensions similar to those of the Earth but greater mass (5 Earth masses) and orbits in the habitable zone of a star very similar to the Sun. Problems for life could be linked to the age of the star, which, presumably radiating approximately 10% more energy than the sun does, could have triggered a growing uncontrolled greenhouse effect similar to that which can be detected on Venus in the solar system. Researchers from the *SETI Institute* (Search for Extra-Terrestrial Intelligence) are using a radio telescope in California to search for radio transmissions from Kepler-452 b. *Kepler-1638 b* (ESI 0.76) also orbits a star similar to the Sun in terms of *mass, temperature, age, and metallicity*. In addition to stars similar to our Sun, there are slightly cooler stars, *orange dwarfs* that remain stable for much longer than *yellow dwarfs* such as the Sun, which is why they are often indicated as the best candidates around which planets could exist habitable. A planet with a mass a couple of times greater than that of Earth is *Kepler-442 b* and has a habitability index even higher than that of Earth. *Kepler-62 e* (ESI 0.83) also orbits an orange dwarf. The planet, with a radius slightly larger than that of the Earth, is probably a super-Earth with a solid surface and is located in the habitable zone of the star, where the presence of liquid water on the surface is possible. Other planets revolving around orange dwarfs include *Kepler 1544b* and *Kepler 283 c*. There are also planets in multiple systems, such as *Gliese 667 Cf* (ESI 0.76) and *Gliese 667 Ce* (ESI 0.60), which revolve around Gliese 667, a multiple star system made up of two orange dwarfs, a little cooler than the Sun, and a red dwarf. *Kepler-296 e* (ESI 0.85) is one of the 5 planets that revolves around the binary star Kepler 296, which is made up of an orange dwarf and a red dwarf. In the list of planets we have seen, there are several that could host life, and among these, we remember *Teegarden b, Trappist-1 e, Kepler-442 b, Kepler-452 b, Kepler-1649c, Ross 128b and K2-18b*. Clearly, being habitable does not imply that there is life on the planet. Further studies of the atmosphere are needed in search of the chemicals of interest. **The interested reader can have a look to Appendix C to have a deeper insight on abitable planets**.

8.6   The Traces of Life

In Chap. 6, we observed that many conditions must be met for a planet to host life. It is not enough just that it is in the habitable zone. With respect to the Earth-like planets in the habitable zone, we had to speculate on whether there could be liquid water and, in general, whether there could be conditions for life. The question we can ask ourselves is whether there is a precise way to establish whether, looking at it from afar, there is life on a planet. One possibility is to study the light passing through their atmospheres with the goal of finding a compound that is related to life or is the product of life. In astrobiology, any substance that provides scientific evidence of the presence of life is called a *biosignature*. What are the most important biosignatures?

Carbon dioxide is emitted by most living things, but there are natural processes that produce it. Therefore, its presence tells us that there is an atmosphere on a planet. *Water vapor* may provide more precise indications. *Water* is certainly a necessary but not sufficient condition for life. Its presence tells us whether a planet can be habitable. *Ammonia* and *nitrogen oxide* contain nitrogen generated by biological processes. These two gases can be generated by natural processes but not in large quantities. Therefore, the presence of the two biosignature gases in an atmosphere is a strong signal, although not definitive, of the presence of life. *Methane* consists of four hydrogen atoms and one carbon atom. It is produced not only by bacteria in the intestines of animals but also by natural processes. However, *if there were methane, ammonia and nitrous oxide in the atmosphere, we would have a clear indication of the existence of life.* The only way to produce oxygen and keep it constant in the atmosphere is through photosynthesis. The presence of oxygen in the atmosphere is an indicator of the presence of life. At the same time, its absence would not indicate the absence of life. On Earth, for billions of years, there have been life forms capable of living without oxygen. *Ozone* is also an important biosignature and protects organisms from ultraviolet radiation. Another key biosignature is *chlorophyll,* which produces carbohydrates from carbon dioxide and water. It also produces oxygen as a byproduct. Detecting it is difficult via spectroscopy, and the only sign of its presence is a greater tendency toward reddish coloration of the light reflected from the planet. In addition to these main biosignatures, there are many others that specialists know very well. For atmospheres such as Earth, *methanothiol, chloromethane* and other *sulfur-containing gases* have been proposed. In hydrogen-dominated atmospheres, the biosignatures include *methyl chloride, dimethyl sulfide* and *nitrous oxide.* There are three different methods used to study biosignatures. The first is that of transmission spectroscopy. When the planet passes in front of the star, the star's light diminishes, and gases in the planet's atmosphere also absorb some wavelengths. As previously mentioned, when light passes through a gas, an absorption spectrum of lines that correspond to the elements that produce the absorption (Fig. 8.2) occurs. Dozens of hot Jupiter atmospheres have been studied. Now, we should do the same with terrestrial-type planets.

The second is *reflectance spectroscopy.* When a planet passes behind a star, the star's light can bounce from the atmosphere and reflect back to Earth. The third is *emission spectroscopy.* When a planet is in front of a star, if it is very hot, it can emit radiation that can be observed and studied. In February 2016, the Hubble Space Telescope was used to study the atmosphere of the Super-Earth 55 Cancri. No water was detected, but many molecular hydrogen, helium and traces of cyanide were detected. These findings have led us to believe that

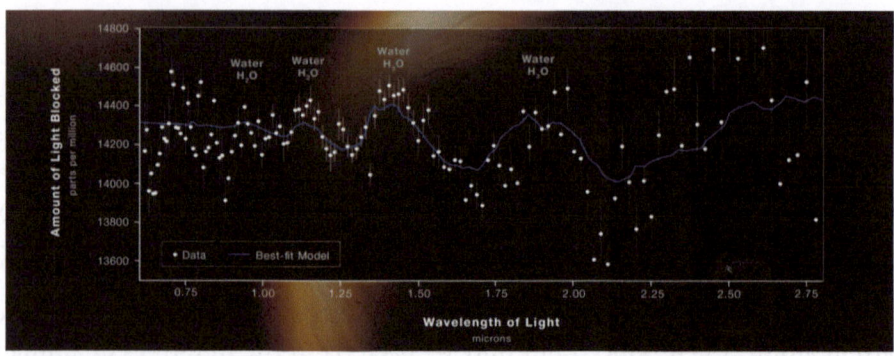

Fig. 8.2 Spectrum of WASP-96 b obtained with reflectance spectroscopy through James Webb Telescope. (Credits: NASA)

the planet is a "diamond world" very rich in carbon and an abundance of diamond within it. It is easier to observe biosignatures via infrared or millimeter waves, but very powerful telescopes are needed. James Webb is capable of looking for biosignatures. It was used to determine the spectrum of the gas planet *WASP-96b*. The spectrum revealed the presence of water. James Webb will be used to study *TRAPPIST-1e*, which we discussed earlier, with transmission spectroscopy. It is able to detect changes in atmospheric levels of carbon dioxide, water vapor and methane, and it is not designed to do more complicated tasks. For example, it cannot detect the presence of oxygen. There are plans for space telescopes that use reflection spectroscopy and block light with coronagraphs to reveal the light reflected from the planet. One of the most ambitious projects is the *ATLAST* (Advanced Technology Large-Aperture Space Telescope), a space telescope that could have a diameter of 16 m, even larger than the James Webb (6 m), allowing astronomers to answer cutting-edge questions of modern astrophysics, such as that of the existence of life in the galaxy. Ground-based telescopes such as the *E-ELT* and the *TMT* (Thirty Meter Telescope) will also be used. In 2015, NASA created the *NExSS* program, a coordination network intended to study planetary habitability. With their help, the atmosphere could be probed for free oxygen on extrasolar planets. Let us also remember the project *Breakthrough Starshot* mentioned in the introduction, which could bring small sailing ships to Proxima Centauri.

9

Carbon Chauvinism?

Since the heliocentric hypothesis has triumphed over the geocentric hypothesis, the *Copernican principle* has prevailed, namely, the simple fact that we have nothing special. If this principle was a universal principle, of course, then we could think that the Universe is full of earths and that these earths are inhabited by living beings. Now, the results of astronomical research in recent decades tell us that the Universe is actually full of planets, that among them, in a minority, there are earths, but the conclusion that there are living beings on them, perhaps evolved as more than us, rests on nothing. To reach this conclusion, we would have to demonstrate in the laboratory that life originates easily, but we have not succeeded in doing that or find planets on which life exists. Today, our position can oscillate between those of de Duve and Monod without the risk of being wrong. The origin of life is possible first if there is an energy source, i.e., an environment out of thermodynamic equilibrium. Life on Earth is due to sunlight, which is captured by living forms, gives rise to a chemical imbalance in those living forms and which gives rise to the food chain. However, there are organisms that can also live in a strong lack of solar energy, such as in the black fumarole environments that we described in Chap. 4. Energy linked to volcanism or plate tectonics is another source of energy for life. When there is a lack of balance, this generates a flow of energy that living beings skillfully use and dissipate. The complexity of life on Earth is based on reproduction, which triggered evolution by natural selection. On this basis, there is therefore the existence of individuals who interact with each other, and individuals exist because cellularization has developed during evolution, i.e., the individual is composed of cells. The life that exists on Earth is based on these two pillars: the existence of thermodynamic imbalances and cellularization, and the element on which everything rests is the chemistry of carbon. Assuming that life exists in the Universe, one might ask whether there could be life that is based on other pillars, namely, life that is not based on carbon and water. Life requires elements that can give rise to large molecules, some of which are capable of storing information, such as RNA and DNA. The first condition for life beyond carbon is that the element on which life is based is abundant enough and capable of creating complex and stable molecules. There are 91 elements present in nature, and those that could play a role in the formation of life must be stable and able to form at least three *covalent bonds*, i.e., chemical bonds in which two atoms share pairs of electrons. Two bonds are necessary for the element to form bonds with itself and other elements and form long chains or rings. The remaining bonds serve to bind with other elements and create structures capable of conveying information. The elements that meet these requirements are as follows: the elements that form three bonds: boron, nitrogen, phosphorus, arsenic, and antimony; and the elements that form four bonds: carbon, silicon, germanium, and tin. As mentioned, a good quantity is also needed, and in the Universe, the most abundant among these are carbon, nitrogen and silicon. Among these three, carbon is the one that has the greatest ability to form covalent bonds with itself and with other elements and form stable and long chains (e.g., proteins and nucleic acids). Furthermore, nucleic acids have a sort of skeleton with negative charges, which, by repelling, keep them stretched out so that the change in the nucleotides does not affect the structure of the DNA. Instead, proteins have no charge, so they can fold. These characteristics should also be respected by a type of life on the basis of elements other than carbon.

9.1   Silicon Life?

The element most often cited as a possible basis for noncarbon life is *silicon*. In the periodic table, it is found under carbon, and therefore, in the external layer, a silicon atom, like carbon, has 4 electrons, which intervene in chemical reactions. Silicon is less abundant in the Universe than carbon is but more abundant in the Earth's crust and is abundant enough for life to form on the basis of it. Like methane CH_4, silicon can combine with four hydrogen atoms to form *silane* SiH_4. Having said that, we must remember the negative sides of silicon for the purposes of a life on the basis of it. First, the bond between silicon and hydrogen is much more reactive than the carbon–hydrogen bond is, and therefore, silane is less stable than methane. When more carbon and hydrogen atoms join together, hydrocarbons are formed, whereas more silicon and hydrogen atoms form silanes and *polysilanes*, which are less stable than *hydrocarbons* are. Therefore, while long chains of carbon atoms are easy to form, they are not the same as silicon, and polysilanes are scarce in nature. Furthermore, silicon compounds ignite in the presence of air; therefore, silicon-based biochemistry requires an oxygen-free environment and solvents other than water, such as methane, nitrogen or ethane. Under these conditions, polysilanes are nucleic acids in the silicon world. Silicon can form main chains with oxygen and side chains with carbon. These compounds are called *siloxanes* and are stable and used in many everyday products (food additives, insulators, cosmetics, silicones, etc.). It is unclear whether these compounds can support life. However, in general, a world with silicon life should be very cold (since solvents such as methane are needed), and there should be no water or oxygen. While carbon compounds are found everywhere in the cosmos, silicon compounds are rare in meteorites. For the reasons mentioned and various others, silicon does not seem like a candidate that could replace carbon in the formation of life; at most, it could collaborate with it. In fact, many probiotic reactions are facilitated by the presence of clay (made up of silicates) in an aqueous environment. As seen in Chap. 4, Jack Szostak, in 2001, accelerated the speed of vesicle creation by a factor of 100, leading to the formation of cells by adding small amounts of a kind of clay in his experiments on the origin of life on Earth. Other researchers, such as Cairns-Smith and Martin Brasier, have also suggested the importance of clays and other silicates for the origin of life.

9.2 Which Solvents?

The other aspect of great importance is determining the best solvent for life. The chemical reactions necessary for life to occur more easily in a liquid state. Water is very effective as a solvent, but in the case of the comparison we made between carbon and silicon, we can see if it can be replaced by another solvent. The liquid base is essential for nutrient transport. At typical Earth temperatures, the best solvents are *water, formamide* and *sulfuric acid*, whereas at low temperatures, the best solvents are *methane, ethane, ammonia and liquid nitrogen*. At high temperatures, however, *molten silicates* and *silica exist*. High temperatures, however, destabilize proteins and nucleic acids, and for this reason, it is thought that the limit temperature for life is 150 °C. With respect to water, we know well from our daily experience how fundamental water is for life. Water is composed of a central oxygen atom and two hydrogen atoms located on a plane. The three atoms form a sort of letter V with an angle between the two 105° bonds. It is a polar molecule, meaning that oxygen has a partial negative charge and that hydrogen has a partial positive charge. Polarity allows the formation of bridges (called *hydrogen bonds*) between water molecules and between these molecules and other molecules. These bridges make water a great temperature stabilizer. When water freezes and forms ice, it expands, unlike most other substances. Owing to this aspect, ice floats, making it difficult for large volumes of water to freeze. A negative aspect is the fact that ice strongly reflects solar rays, producing cooling of the surrounding regions and, under certain conditions, leading to the phenomenon of *glaciation*. Water, as a polar molecule, favors the grouping of nonpolar substances in an aqueous environment, which is important because the nonpolar part of the phospholipids clumps together, forming important membranes to isolate cells from the outside. The same phenomenon occurs in proteins within which amino acids take refuge. On early Earth, water had a negative effect. This process results in breakdown reactions that hinder the formation of nucleic acids and proteins. Today, this trend of water is a positive aspect of obtaining metabolic energy. Another solvent, similar to water, is *ammonia* and is quite abundant in the Solar System. It forms bridges such as water but with greater difficulty. Ammonia dissolves many organic compounds and hydrophobic (water-averse) molecules better than water does, but this is not always an advantage. Other solvents are hydrocarbons such as methane and ethane, which have the advantage of not destroying compounds by splitting them into two or more parts. They are abundant in some places in the solar system, such as Titan. In a lake or sea, there could be conditions for the formation of vesicles, which then form cells made of nitrogen, carbon and hydrogen. *Sulfuric acid* is another possible solvent that could support life, although it is corrosive in the presence of water. *Formamide* behaves like water but is less reactive. Importantly, important compounds such as peptides are formed and can serve as precursors in the formation of RNA. *Silica* is a candidate perfect solvent, but life is unlikely to thrive at temperatures above 1700 °C. In conclusion, water remains the most suitable solvent, among other things, because it is the most abundant triatomic molecule. One of NASA's mottos, to search for life, is precisely the principle of *following the water*. For life, in addition to the need for a solvent and a biogenic element, a form of usable energy is needed. On Earth, the most important energy sources are sunlight and chemical energy. The first is used by surface organisms, and the second is used by organisms in the deep sea, such as black fumaroles, in which there are large thermal imbalances and differences in the concentrations of different compounds used by the organisms that live there. Alternative forms of energy could be pressure differences between different points on planets such as

Venus and the gas giants. In solar system satellites with underground oceans, energy may come from tides. Another important aspect is whether there may be paths to life other than those that exist on Earth. As we mentioned in Chap. 4, life, as we know, is based on three aspects: *metabolism, genetics* (RNA, DNA) and *cellularity*, and the formation of cells that separate the individual from the outside world. Now, as discussed in Chap. 4, we cannot understand how we can go from RNA to the DNA–RNA–protein system. Could it be that life remains confined to RNA or RNA precursors? Giving an answer to this question is perhaps impossible because we only know about terrestrial life, and somehow, we do not understand how nature managed to move from RNA to DNA and combine the two to form proteins.

9.3 Universal Convergences

From an opposite perspective, one could instead think that there are some sorts of *universal convergences* in evolution and that if different life forms exist on other planets, these could have characteristics similar to those of terrestrial life. A study of the *Ediacara fossil* deposit in Australia dating back to the *Cambrian explosion*, which occurred 570 million years ago, when the majority of complex animal types appeared, of which only a few survived, clearly revealed that many multicellular morphologies were explored on Earth. This leads us to think that there are universal morphological patterns because they are advantageous. The formation of vesicles, which are the basis of cellular formation, from the self-organization of *amphipathic* compounds, i.e., molecules in which polar and nonpolar groups coexist, is not a coincidence. This is supposed to happen everywhere in the Universe. Similarly, the fact that certain symmetries in living beings predominate over others is because they are more stable and easier to generate. In 2015, Simon Conway Morris, in his book *The Runes of Evolution,* argued that animal and plant life on other planets, if it exists, must be similar to that on Earth. One of the examples he gives to reach these conclusions is that although octopuses and vertebrates appeared independently, their eyes are very similar. The shapes and fins of fish, to move in water, such as the wings of birds, have been "reinvented" many times so much that they can be considered universal shapes. The *echolocation* used by birds to orient themselves in the air has also been reinvented in the case of cetaceans to orient themselves in the aquatic medium. These are just a few of the hundreds of examples of evolutionary convergences that Morris discusses in his book. In other words, evolution, as a type of experiment performed on living forms, always has the same patterns. For this reason, according to

Conway, extraterrestrial life, if it exists, cannot be so different from what we observe on Earth. Gravity and the type of star must also be considered. In planets with stronger gravity, a presumed animal is smaller in size and has a more robust body structure. The color of plants depends on the dominant wavelength emitted by the stars. Astrobiologist Dirk Schulze-Makuck claimed that, for example, on Titan, which has a lower surface tension than water does, bacteria are larger, and the types of life present on a planet depend on how much energy is available. Thus, on the seabed of Europe, there could be no more complex beings than shrimp. All these speculations are clearly based on the hypothesis that extraterrestrial life forms exist. However, some researchers believe that the Earth is a unique case in the Universe, the so-called *rare Earth hypothesis*. Two of the scientists who have most strongly defended this idea are Peter Ward and Donald E. Brownlee. In their 2000 book, *Why Complex Life is Rare* highlights that the condition of the Earth is very particular. From its position in the galaxy, the characteristics of our star, its position in the habitable zone, the magnetic field, the existence of a moon with dimensions that it does not have similar dimensions in our solar system and that would be important for the stability of the axis terrestrial, to the existence of Jupiter, which would act as a protector from asteroids and comets.

In conclusion, it seems that the life forms that could exist in the Universe should be based, like we, on carbon, that water is the most abundant and good solvent, but there could be worlds in which it is replaced by other solvents, such as ethane or methane (case of Titan).

10

Where Is Everybody?

One summer day in 1950 while working at laboratories in Los Alamos, New Mexico the scientist Enrico Fermi went to lunch with colleagues, including Edward Teller, Herbert York and Emil Konopinski. The latter mentioned a satirical magazine cartoon showing aliens stealing garbage cans in New York City. They first began to discuss intelligent life in the universe, superluminal travel, and other topics. After a long period, during lunch, Fermi pronounced: *Where is everyone*? obviously referring to aliens. Furthermore, according to York, Fermi began to perform several calculations, and on the basis of these calculations, he concluded "…that we should have been visited a long time ago and many times".

The previous question encapsulates what is now known as the *Fermi paradox*.

The logic in the question was as follows: given the antiquity and vastness of the Universe, the enormous number of galaxies and stars, civilizations older and more developed than ours, should have colonized the galaxy. Therefore, in a population almost 14 billion years old and in a galaxy of hundreds of billions of stars, the question naturally arises as to where the aliens are. Since there is no concrete evidence of their existence, Fermi and his paradox concluded that they do not exist. This argument convinced many scientists who devoted themselves to everything other than the problem of life outside the Earth. However, there is a small group of scientists who have a different point of view than those of Fermi.

A. Del Popolo, *Extraterrestrial Life*, https://doi.org/10.1007/978-3-031-83497-4_10

10.1 Technical Tests for the Search for Extraterrestrials

One of these was Frank Drake, a doctoral candidate in astronomy at Harvard University. In 1957, working on his thesis, Drake was looking at the seven Sisters. The Seven Sisters are the Pleiades in astronomy. In the literature, there are traces of them in the Chinese annals of 2350 B.C. and in Hesiod's poem from 1000 B.C. in the Iliad and the Odyssey. Sailors refer to this group of stars for navigation and farmers for harvests. In Greek mythology, the Pleiades were seven sisters: Maia, Alcyone, Asterope, Celeno, Taygete, Electra and Merope. Daughters of Atlas, the titan to whom Zeus had entrusted the task of supporting the Earth, and of Pleione, the protector goddess of sailors. Following a fortuitous encounter with Orion, the Pleiades and their mother became the hunter's prey. To protect them from their nagging amorous assaults, Zeus transformed them into her doves, releasing them into the sky. Zeus is also said to have fathered three of the sisters. There are Native American legends, Aboriginal legends, and Hindu legends about these stars. In Japan, the seven Sisters are known by the term "Subaru", which in Japanese means "united" and "unity". When the car company of the same name chose the name Subaru, it decided to reproduce only six of the seven stars in its logo because these are the only ones that are actually visible to the naked eye. Drake was studying the abundance of hydrogen in stars with the aim of understanding how they are born. One February night, while stargazing, he flashed a signal on the radio telescope's screen. Drake thought that the signal was not of natural origin and that the idea ran through his mind that it could be a signal sent by extraterrestrial civilizations. The signal remained there, and it did not disappear. Then, Drake moved the antenna and observed that the signal did not disappear; therefore, it must be of terrestrial origin. In the following weeks, Drake began to mull over the idea of the existence of alien life forms sending signals into space and thought that humanity should also take it upon itself to look for those signals and perhaps make contact with those civilizations. After completing his doctorate, he moved to Green Bank, West Virginia, where some radio telescopes are located. In addition to Drake, another young scientist, Philip Morrison, who had obtained his doctorate in Berkley, California, began to think that aliens could send us messages. Together with Giuseppe Cocconi, in 1959, he published an article in Nature. In this article, the two argued that an advanced extraterrestrial civilization, having understood that there was intelligent life on Earth, could have started sending signals toward us, hoping to have an answer. On what frequency should we look for this signal? Since the most abundant element in the Universe is hydrogen, and this element emits radiation at a frequency of 1420 megahertz corresponding to a wavelength of 21 cm, the two indicate that this wavelength is ideal. This frequency is distinct from the band in which most of the cosmic background radiation signal is found. The Green Bank is an isolated region with very few television and radio stations and has little interference from human activity. Drake obtained from the director, with whom he had become on good terms, time for observations and the search for extraterrestrial signals. Named after the Queen of Oz, the project was named *Project Ozma*, which he began in 1960. First, the search begins in the direction of the star *Tau Ceti,* and then, the antenna is directed to *Epsilon Eridani.* From this observation, he received a pulsating signal that disappeared shortly after. In that area, the military carried out experiments that most likely generated the signal received by Drake. Drake did not give up and continued to observe other stars, to no avail. The problem that arose was which of the enormous number of frequencies should we look for. Drake and colleagues followed Morrison and Cocconi's reasoning and began looking for signals in the microwave range, including the frequency indicated by Morrison and Cocconi.

10.2 Civilizations in the Universe?

In 1961, the National Academy of Science of the United States organized a conference at the Green Bank Observatory on the search for signs of extraterrestrial civilizations. Drake was designated an organizer of the conference and thought that if the conference went well, it would bring funds to Project Ozma and the search for alien signals. After much thought, Drake brought a single formula to the agenda of the conference. The latter is now known as the Drake equation. As Drake wrote,

As I planned the meeting, I realized a few day[s] ahead of time we needed an agenda. In addition, so I wrote down all the things you needed to know to predict how hard it is going to be to detect extraterrestrial life. In addition, looking at them, it became pretty evident that if you multiplied all these together, you got a number, N, which is the number of detectable civilizations in our galaxy. This was aimed at the radio search and not at searching for primordial or primitive life forms.

The equation was composed of the product of various terms

$$N = R^* \, f_p \, n_e \, f_l \, f_i \, f_c \, L$$

where

- N is the number of extraterrestrial civilizations present today in our Galaxy with which we can think of establishing communication;
- R^* is the average annual rate at which new stars form in our galaxy;
- f_p is the fraction of stars that have planets;
- n_e is the average number of planets capable of hosting life forms;
- f_l is the fraction of planets on which life has actually developed;
- f_i is the fraction of planets on which intelligent beings have evolved;
- f_c is the fraction of extraterrestrial civilizations that have developed a technology;
- L is the time frame in which civilizations can transmit signals that can be picked up on Earth (Fig. 10.1).

To know the number of extraterrestrial civilizations, all the terms of the equation need to be known. Drake and collaborators provided estimates of various parameters. $R^* = 10$ stars per year, $f_p = 0.5$, assuming that half of the stars have planets, $n_e = 2$ (i.e., each planetary system has two planets that can support life), $f_l = 1$, f_i and $f_c = 0.01$, taking the Earth as a model, $L = 10,000$. This yields a value of $N = 10$. In Drake's time, it was not even known if extrasolar planets existed. The most recent discoveries have led to new estimates for various parameters, especially astrophysical ones. In the Drake equation, the

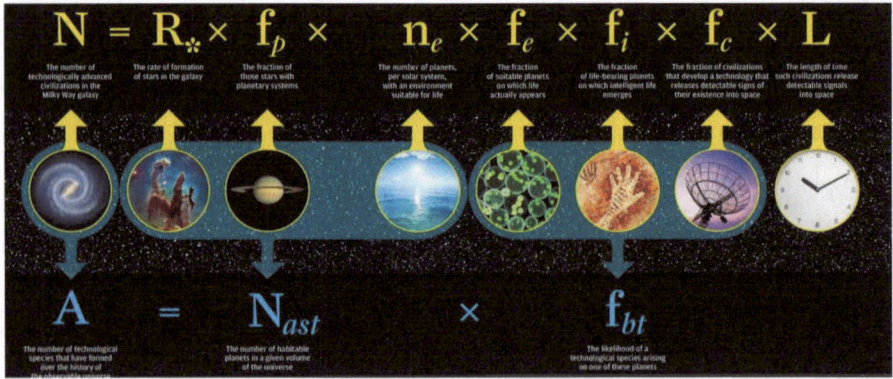

Fig. 10.1 Drake's equation. (Credits: University of Rochester)

terms R^*, f_p, and n_e are obtained from astrophysics and are quite well known today. The terms f_l, f_i and f_c are related to biology and are therefore almost unknown. The last term is linked to the lifespan of an advanced civilization, and we still do not know how much it could be worth. Therefore, even if much progress has been made from 1961 to today in the determination of the parameters of astrophysical origin in the Drake equation, the lack of knowledge of f_l, f_i and f_c and L does not allow us to have a precise idea of how many civilizations there can be in our galaxy or in the Universe. However, it is interesting to review the modern estimates of the astrophysical parameters and review the discussions that have taken place to determine the possible values of the other parameters.

• Let us start with the first term R^*.

This term is one of the best known terms in the equation. Calculations by the ESA and NASA in 2006 suggested that the star formation rate was 7 stars per year, whereas subsequent calculations, in 2010, led to a lower value: 1.5–3 stars per year. As discussed in Chap. 6, not all stars are suited to having planets that support life. Which stars are best suited to life? As we have seen, stars O, B, A and F3 are too energetic and not long-lived; thus, they are not suitable for supporting life. The stars from F4 onward, together with the G and K stars, seem the most suitable to support life owing to their stability, longevity, and duration of the habitable zone. In particular, K stars (orange dwarfs, 5% of all stars) are considered the most suitable for life. Since these stars are less luminous than the Sun is, their habitable zone is closer to the star. M-type stars are also interesting. They are the most abundant and have a very long life. They suffer from two problems: in their first billion years, they are very active, and this is not good for life. As they are dimmer than F-, G-, and K-type stars are, habitable planets are located close to the star and usually always have the same face as the star. On these planets, there would be only a narrow region where life would be possible.

• Therefore, we have the factor f_p, the fraction of stars that have planets.

Since the discovery of exoplanets, we have known approximately 5000 planets and have estimated the possible number of planets in the galaxy and in the Universe. For our galaxy, a 2012 study in Nature by Cassan and collaborators estimated one or more planets per star, and recent studies confirmed these numbers. Therefore, in our Universe and in our galaxy, there is certainly no shortage of planets, both those not suitable for life and those suitable for life. Therefore, the value of the factor f_p can be set equal to 1.

- The next factor in the equation is n_e, the number of habitable planets.

The first thing we need to ask ourselves is what conditions lead to life. For life, a solvent is necessary within which molecules can move and form complex organic compounds for the formation of proteins. Various solvents are possible, but we have seen that water is perhaps the best. Water must be in a liquid state because chemical reactions are much easier in that state. Therefore, the temperature must be adequate for the water to be in a liquid state, which occurs in the habitable zone. The very definition of a habitable zone *is the area around a star in which water is found in a liquid state.* The habitable zone around the Sun is located in the region 0.95–1.37 astronomical units from the Sun. Below 0.95 astronomical units, water evaporates; at distances greater than 1.37 astronomical units, it tends to freeze. From what we observed in Chap. 5, it is possible that life exists on some satellites of the giant planets. These are located at a distance at which water is generally not liquid, but for particular conditions, such as the effect of the tidal forces of the giant planets, it can be in the liquid state in an underground ocean, or it can have a liquid solvent, other than water, on the surface, as in the case of Titan, on which methane flows freely and forms lakes and seas. Therefore, the definition of a habitable zone must be extended and modified. The fact that a planet is in the habitable zone does not guarantee that life will arise on it. An example is Mars. Other conditions are needed: the right eccentricity, rotation period, and mass. In addition to the solvent, a basic element is needed to which the other elements bind to form organic compounds, DNA, proteins, etc. Life, as we know it is based on carbon and as discussed in Chap. 9, even if other elements are proposed, carbon seems to be the optimal solution. We therefore need an energy source, which can be the light that comes from a star. We have seen that there is probably one planet per star. How many of these planets are habitable? According to a 2020 study by Michelle Kunimoto and Jaymie M. Matthews, in our galaxy, there should be 6 billion rocky planets with dimensions similar to those of Earth that revolve around stars similar to the Sun. This number assumes that in our galaxy, there are 400 billion stars, of which 7% are stars similar to our sun, and for each of these stars, there would be 0.18 Earth-like planets, i.e., approximately 5 billion planets (slightly less than the 6 billion indicated) around stars similar to the Sun and located in the star's habitable zone. Considering only the stars similar to the Sun, the product $f_p\, n_e$ is approximately 0.015, but we must take into account that the stellar systems that can have habitable zones are not limited only to stars such as the Sun but also to red dwarfs, as we observed in Chap. 8. Estimates from 2013 revealed approximately 40 billion terrestrial planets in the habitable zone of stars, similar to the Sun and red dwarfs. In this case, the product $f_p\, n_e$ is approximately 0.1. To these, we must add the possibility that there are moons of extrasolar gas giants that could host life, such as some satellites of giant planets (e.g., Europa, Titan, and Enceladus) in our solar system. Hence, the value of the product $f_p\, n_e$ is most likely closer to 0.1 than 0.01. Therefore, with the value of f_p set to 1, n_e should have a value of 0.1–0.2.

- The next parameter is the fraction of planets on which life has actually developed.

The value of this parameter is extremely difficult to estimate since we have only Earth-related data. Estimates cannot be made on a statistical basis. From an observational point of view, information about f_l can be obtained from the study of the alterations induced by life forms on the chemical composition of the planet that hosts them. In fact, living beings release methane and oxygen into the atmosphere. The presence of oxygen and methane is an unmistakable sign of the existence of life forms on a planet. We therefore need tools that allow us to collect a sufficient quantity of the light emitted by exoplanets and carry out a spectral analysis of the atmosphere. Until we do not have concrete data, it will be difficult to understand whether life exists on a planet or not and to obtain an observational value of the term f_l. From the knowledge about life on Earth, which seems to have appeared very early, as soon as favorable conditions arose, it could be deduced that the formation of life is a common phenomenon. If evidence of life was found on Mars or on the satellites of the giant planets and evidence that such life arose independently, then one could conclude that f_l is close to 1. If life on Earth had begun more than once, this would always favor a high value of f_l, but we have no proof of this. In 2020, Tom Westby and Christopher Conselice, researchers from the University of Nottingham, proposed the *astrobiological Copernican principle*, which concludes that life and even intelligent life is formed as a direct consequence of evolution; therefore, in a few billion years, we expect that in habitable extrasolar planets, life forms evolve. From this conclusion, it follows that f_l, f_i, and f_c must all equal 1. Their conclusions are that there are more than thirty civilizations in our galaxy. Ultimately, aside from the argument of Westby and Conselice, we can conclude that f_l is 1, but it could also be much smaller.

- What about the parameter f_i, i.e., the fraction of planets on which intelligent beings have evolved?

Even for this parameter, there are no solid bases for fixing its value. We cannot go much further than conjecture, given that the only example of a planet on which intelligent life has formed (f_i) capable of developing technology and communicating (f_c) is the Earth. We can also try to understand from studies performed on Earth whether intelligence appears naturally. From fossil records, it can be deduced that the evolutionary process on Earth started slowly and then accelerated. For more than half of Earth's life, life has not gone beyond the form of prokaryotic cells without a nucleus. Only two billion years ago, cells acquired a nucleus and began to work together, giving rise to multicellular organisms. Until some time, it was thought that the most complex life forms appeared half a billion years ago with the Cambrian explosion: in a period between 70 and 80 million years, almost all animal groups developed. It actually appears that before the Cambrian, there was a large volcanic eruption in China. Studies of this eruption have led to a changing understanding of the development of life on Earth. Traces of ancient marine organisms, which are more complex than sponges and jellyfish and preserved in phosphates, have been found. These organisms have bilateral symmetry. These findings led us to rethink the traditional theory according to which the great diversity of life on Earth appeared during the Cambrian explosion. These findings were dated to approximately 600 million years ago, and it is now thought that the first multicellular organisms may have evolved 500 million years before the Cambrian explosion. Furthermore, as we have already stated in Chap. 4, the uniformity of life on Earth, as seen at the microbiological level, leads to the conclusion that all organisms have descended from a single cell (the LUCA). How did this evolution, which led from unicellular beings to humans, happen? Unicellular organisms are already at a higher level than viruses: they can respond to chemical changes in the environment, react to light, look for food, and try to reproduce. In 1903, Robert Falcon Scott discovered a region in Antarctica that was extremely desolate, so much so that he called it the "*death valley*". The region was better studied in the subsequent *Terra Nova Expedition* between 1910 and 1913, again led by Scott, who named it *Taylor Valley* in honor of a geologist from the expedition. New explorations revealed large numbers of microscopic plants, single-celled animals, and worms called *nematodes* in the region. In this ecosystem, the bacteria ate the algae, and the nematodes ate the bacteria. Taylor Valley provided important information on the development of the first multicellular organisms. Three and a half billion years ago, this ecosystem was the only one on Earth: the carbon was taken up by algae and then passed on to bacteria and nematode worms. Nerve cells appeared from multicellular organisms. These branches are made up of a nucleus and branches called *dendrites*, which receive information. Dendrites connect to the *axons* (a long filament along which a neuron transmits information) of other nerve cells. Each nerve cell integrates information from neighboring cells and then sends a signal along the axon. The first nerve cells develop into sense organs such as the eyes, nose, and ears.

The collection of nerve cells generated the brain. As organisms evolve, the brain becomes increasingly complex. Charles Darwin realized that there was a correlation between the relative size of the brain and intelligence. Today, we discuss the *cephalization index*, i.e., the ratio of the size of the brain to that of the body. Every living being has its own cephalization index. The highest is that of mammals and among those of primates. Interestingly, cephalization increased over time. Therefore, while between 60 and 40 million years of age, animals had a small brain, this doubled in the period between 55 and 25 million years. Although scientists argue that intelligence is a fortuitous product of evolution, the study of the evolution of the brain and intelligence on Earth disproves this view. Saurians, such as lizards, geckos, lizards, and iguanas, which live on islands, are exposed to new dangers when predators are introduced. In a few generations, they learned to climb trees to escape predators. In other words, "intelligence" can develop over short periods. Other experiments carried out on rats, described by S. Dehaene in *The Number Sense*, showed that rats can count and, according to Dehaene, our ability to understand mathematics derives from the sense of numbers. Lower forms of intelligence have also developed in other animals. The discoveries of recent years provide support for the idea that the evolution of intelligence is something intrinsic to life and that it must manifest itself in sufficient time. Intelligence has evolved in different branches of the animal kingdom. This finding indicates that intelligence is not a coincidence but rather a natural outcome of the evolution of living systems. Evidence for the separate evolution of intelligence in various branches of living systems comes from the theory of bird evolution. Birds are different from other animals, and it was not clear what they came from. Long studies have highlighted that they descend from dinosaurs, and their brains have developed much more than those of fish and reptiles. Their levels of cephalization are much closer to those of mammals than to those of reptiles. Ultimately, we do not take anything away from man if we say that other living creatures have their own intelligence, albeit inferior. The first multicellular organisms appeared approximately 600 million years ago, and the first brains, collections of nerve cells, appeared around the same time. Mammals appeared approximately 200 million years ago, and primates appeared approximately 60 million years ago. Our ancestors appeared only 4 million years ago. *Homo sapiens* appeared between 200,000 and 130,000 years ago. This lineup tells us two fundamental things: evolution times are very long, and evolutionary development has always been moving toward constantly growing intelligence. On the basis of these ideas, several scientists give f_i a value close to 1. However, opposing points of view exist. Proponents of the *rare Earth hypothesis* assume a particularly low value, and biologist Ernst Mayr agrees with this. His point is the opposite of what we discussed before and which shows the history of evolution. According to him, there are billions of living or extinct species on Earth, but only one shows true intelligence; therefore, f_i would have very little value. Obviously, Mayr refers to human intelligence, but it is clear that there is not just one intelligent species on Earth. A study by David Kipping of Columbia University concluded that if the evolutionary clock were to be restarted from scratch, intelligent life would probably not reappear on Earth. Given that the appearance of intelligent life took a very long time, 4.6 billion years, Pascal Lee of the SETI Institute gives a low value, 0.0002, to f_i. I believe that, in this dispute, those scientists are right who think that intelligence grows over time and that species such as man are destined to appear, if there is enough time and that, therefore, considering billions of years of evolution, it must be close to 1.

- There still remains f_c, the fraction of extraterrestrial civilizations that have developed technology and can communicate with other planets.

Because we have not received signals from extraterrestrial civilizations, we have no data for a statistical estimate of this parameter. We must also consider that there may be advanced civilizations that do not use radio wave transmission. Transmission within the community could occur via undetectable signals, for example, through the use of optical fibers. On the other hand, we are also becoming increasingly less visible in the cosmos through radio transmissions, since so many signals are transmitted over the optics fiber. Furthermore, an extraterrestrial civilization could deliberately decide not to transmit radio signals into space. The other problem is that we have limitations in receiving radio signals. For example, if an extraterrestrial civilization sent a signal from 100 light years in all directions, they would have to use a 66,000,000,000 watt antenna because we could receive the signal with the Arecibo radio telescope.

With the *SKA radio telescope* (Square Kilometer Array), which is still under construction and has a size of 1 km^2, the results would be better. To understand the power of this instrument, it will be able to detect the radar of an airport on a planet located 50 years away. Today, we have no better estimates than those of Drake and colleagues in 1961 for f_c. Its value should range between 0.1 and 0.2.

- The last term is the L factor, the lifespan of a civilization.

This factor also has an unknown value; we do not know how long a civilization can last, and as has been noted over the years, it is the most important factor in the equation. This term was introduced into the equation by Drake for the risk of extinction, and it should consider the various elements that influence evolution and the longevity of life, especially intelligent life, at risk. This term should take into account the possibility of self-destruction of an advanced civilization. At the time that Drake wrote the equation, the *Cold War* was in full swing, implying the risk of the destruction of humanity through the use of weapons of mass destruction. Carl Sagan hypothesized that all terms except the lifespan of a civilization are relatively high and that the determining factor in whether a large or small number of civilizations exist in the universe is the lifespan of the civilization, or in other words, the lifespan of a civilization, which in technological civilizations is closely linked to its ability to avoid self-destruction. We live in a "violent" universe with phenomena that produce radiation and release high energy. If a planet is not very far from a star that will explode in a supernova, life on it is at risk. We must not forget the danger from asteroids and comets. In 1994, comet Shoemaker-Levy 9 crashed into Jupiter. The comet broke into fragments when it entered the planet's atmosphere, and these fragments fell onto the surface, generating zones of destruction as large as the Earth. The impact increased the amount of dust, darkening the Jovian atmosphere. Events such as these are not very rare. The extinction of the dinosaurs is attributed to a similar event. This idea is confirmed by the discovery in 1978, in Italy, by Louis and Walter Alvarez of rock layers rich in iridium in the geological layer corresponding to 65 million years ago. Iridium is rare on Earth but is present in meteorites. Continuing their studies, they realized that this layer of iridium was uniformly distributed over the entire Earth's surface and that it must have been produced by the arrival of a 10 km asteroid. In 1991, a large crater was discovered under the Yucatan Peninsula, Mexico. Judging by the size of the crater, collision with the asteroid caused the release of energy equal to that of 5 billion atomic bombs, such as that in Hiroshima. The effect of the asteroid was not only local destruction in the region of fall but also the dust that rose and spread into the Earth's atmosphere blocked the sun's rays for many months, producing a notable reduction in temperature. The absence of light caused the plants and therefore the animals to die. Even today, it is not known which life forms survived and how life was able to recover again. This scenario is similar to what would occur in the case of a global nuclear conflict. On June 30, 1908, a loud bang was heard in the villages of the Tunguska region of Siberia, followed by a ball of fire in the sky. Two thousand square kilometers of forest were destroyed in the event, as was observed 17 years later by the first expedition to the region. This event was most likely generated by a building-sized asteroid. More recently, on the morning of February 15, 2013, in Chelyabinsk, south of the Urals, a fifteen meter meteoroid hit the atmosphere, shattering over the city. The damage caused was not severe. There were injuries from splinters of windows shattered by the shock wave. There are so-called NEOs (near-Earth objects), which are solar system objects that can intersect the Earth's orbit and can produce a collision. Another phenomenon that can be destructive for a planet is volcanism, if excessive, as in the case of Jupiter's satellite, Io. In the case of Earth, the end of the *Minoan civilization* was probably due to the eruption of Santorini volcano in 1646 B.C. This eruption partially devastated the island, called Thera, and it wiped out entire community and agricultural areas on nearby islands and the coasts of Crete. This hypothesis was reported in the article *The Volcanic Destruction of Minoan Crete* published by the English periodical "Antiquity" by Spyridon Marinatos. Returning to the estimate of the value of L, some believe that our civilization cannot survive more than a couple of hundred years of technological development. There are opposing points of view. In theory, our civilization could still exist for a billion years, enough time for the sun to increase its brightness by 10%, and our civilization will not be able to survive. David Grinspoon came to the conclusion that once a civilization had developed enough, it would overcome all threats of extinction and would last for an indefinite period, so according to him, L would be worth billions of years. A completely opposite conclusion comes from the scientific writer Michael Shermer, who, from studying the duration of sixty terrestrial civilizations, came to the conclusion that L must be equal to 420 years, while considering 28 civilizations more recent than the Roman Empire, he concluded that L is equal to 304.

Given our ignorance of the last four factors, the Drake equation can yield results over a very wide range, depending on the assumption. It is possible to obtain values of the number of civilizations N much smaller than one, which implies, given that we are aware of the existence of at least one civilization in our galaxy, namely, ours, that some of the parameters must have a greater value. On the other hand, values of N much greater than 1 can be obtained, which implies the existence of many civilizations. Combining the value of the star formation rate calculated by NASA, $R^* = 1.5$–3 year^{-1}, the low value of the product $f_p \cdot n_e f_l = 10^{-5}$ estimated from the rare earth hypothesis, Mayr's view on intelligence, $f_i = 10^{-9}$, Drake's view on communication, $f_c = 0.2$, and L's estimate of Shermer, 304 years old, we obtain a value of N equal to 9.1×10^{-13}, which would imply that we are alone. Always using the same NASA value for $R^* = 1.5$–3 year^{-1}, different values such as $f_p = 1$, given by J. Palmer in 2012, $n_e = 0.2$, obtained from the results of two articles by various authors published in an important astrophysics journal, $f_l = 0.13$, obtained from C.H. Lineweaver and T.M. Davis, $f_i = 1$, obtained by A. Campbell in 2005, $f_c = 0.2$, given by Drake, and L = 109 years, according to D. Grinspoon, there would be a civilization value equal to 15,600,000, certainly an exaggerated number. In 2009, D. Forgan used a method of simulation of the parameters of the Drake equation on the basis of a model of stellar and planetary distributions, the characteristics of life in the Milky Way and the stochastic nature of evolution and obtained values of N on the order of a hundred. In June 2018, three researchers from the *Future of Humanity Institute* at Oxford University started from the impossibility of obtaining a certain result from unknown variables, repeatedly solving the equation with data taken from scientific publications, random and different each time. According to the average of the results, the galaxy could be populated by one hundred civilizations, but the equation returned the same unfortunate result 30% of the time: zero.

The Drake equation is a simple model that omits many relevant parameters, and modifications to the equation have been proposed. For example, Carl Sagan proposed his own version of the Drake equation, and more recently, Sara Seager proposed another equation based more on the search for planets with gases having biosignatures. These gases are produced by living organisms and can be detected by space telescopes. The equation proposed by Seager is

$$N = N_* \, F_Q \, F_{HZ} \, F_O \, F_L \, F_S$$

where

N = number of planets with detectable signs of life
N_* = number of stars observed
F_Q = number of stable stars
F_{HZ} = fraction of stars with rocky planets in the habitable zone
F_{OR} = fraction of those planets that can be observed
F_L = fraction of planets with the presence of life
F_S = fraction over which life produces detectable gaseous biosignatures

In the end, the Drake equation was modified in 2016 by Adam Frank and Woodruff Sullivan. Instead of asking how many civilizations currently exist, they asked themselves what is the probability that in the entire history of the Universe, only our civilization has appeared. As we will see in detail in Chap. 12, the data for the universe imply that it is extremely unlikely that Earth is home to the only technological species that has ever existed.

What can we conclude from this long discussion? The Drake equation was written only to discuss extraterrestrial life in the 1961 congress; therefore, it has major limitations, and the parameters designed by Drake, particularly the last 4, are not easy to determine. To the best of our knowledge, this equation has been modified to have more easily determinable parameters. The latest results indicate that it is extremely unlikely that a single technological species is present in the galaxy, especially in the Universe. In the next decade or twenty years, with the study of the atmospheres of the discovered and future planets, we will be able to establish whether life exists outside Earth. The Seager equation has parameters that are not difficult to determine. For the moment, we can conclude that it is very unlikely that we are alone.

11

The Great Silence (Searching for ET)

To date, we have no evidence of the existence of intelligence and extraterrestrial life, although the existence of the latter is certainly more probable than intelligent life. We observed in Chap. 9 the existence of *evolutionary convergences*, that is, the tendency of various species living in the same environment to develop, under the pressure of natural selection, certain structures that cause them to resemble each other. We also provide some examples of evolutionary convergences. A question without a certain answer is whether intelligence can be a further convergent character, that is, whether its appearance is only a question of time. In the discussion in the previous chapter, we observed that some scientists believe that intelligence tends to grow over time from simpler to more complex species and that, given enough time, we would reach levels of intelligence similar to those of humans. Others, however, think in a completely different way, such as Ernst Mayr, for whom only one of the 50 billion species that lived on Earth was able to generate civilization and electronic technology. Additionally, for Stephen Jay, the appearance of the human species is due to a succession of contingencies. Some simple estimates conclude that an advanced species could conquer a galaxy in a few tens of millions of years. It is also estimated that habitable earths and super-Earths have existed for at least 9 billion years. Therefore, the question that Fermi posed to his colleagues in 1950 arose spontaneously: *where is everybody?* Although the possibility of real contact between our system and other civilizations may be unlikely, contact via electromagnetic wave transmissions is simpler, which makes great silence rather anomalous. This silence could be due to several reasons: the scarcity of technological life in the Universe or life in general, the great distances of these civilizations, a short life of advanced civilizations that could self-destruct or disappear owing to natural causes. In the latter case, as Aditya Chopra says, the great silence would be because the extraterrestrials are all dead. Another possibility is the lack of interest of advanced civilizations in communicating or that they communicate, but we are not able to recognize such signals because they are sent with technologies we do not know. This discussion leads us to conclude that since we are not certain about the points listed above, we can only try to scan the sky in search of signals or send our own in the hope that some civilizations will receive them and respond to us.

11.1 SETI Projects

Many years have passed since Cocconi and Morrison published their article *Searching for Insterstellar Communications* in Nature in 1959, and Drake began his Ozma project in 1960. What has happened in all these years? Is there any news on the search for extraterrestrial life? As we mentioned when talking about life in the solar system in Chap. 5, several attempts have been made to send probes into our solar system, and Morrison and Cocconi's ideas of looking for signals sent by an extraterrestrial civilization have produced several initiatives. The ideas of the SETI (Search for Extra-Terrestrial Intelligence) project were inherent in the article. To communicate, you need to send and capture a certain form of energy. The best form of energy at large distances is electromagnetic radiation because the Universe is particularly transparent to most of this radiation. The range of possible frequencies is enormous, and the question arises as to which is the best frequency to search. Stars emit very strongly in the visible light range and relatively little in the radio frequency range. When radio waves are emitted from a planet, they are easier to detect. Microwaves in the radio wave band are the best frequencies because there is less natural interference. According to Morrison and Cocconi, there is a frequency of particular interest in this band, which is the emission of hydrogen at 21 cm. Another interesting wavelength is that of 18 cm resulting from the hydroxyl radical (-OH), which originates from the breakdown of the water molecule. The region between 18 and 21 cm, known as the *water hole*, is an interference-free region and is considered an important wavelength region for communication with extraterrestrial civilizations. To be safe, research has also been carried out at many other wavelengths. Two different strategies are usually followed in the search. The first strategy involves turning attention to nearby stars that may have exoplanets, and the other strategy involves scanning the sky over broad wavelength ranges. The sensitivity of current detection systems allows us to identify intentional, i.e., nonnatural, very powerful signals at large distances and distinguish them from signals coming from Earth. Morrison and Cocconi's idea was put into practice in 1960, with the Ozma project, which we discussed in Chap. 10. This project was essentially the first SETI project. Only many years later, in 1984, was an institute dedicated to the search for intelligent extraterrestrial life equipped with technology that would allow it to send signals into the cosmos, called the *SETI Institute*, directed by Frank Drake until his death in 2022. The first conference dedicated to SETI was organized in 1961 in Green Bank, and a few years later, the Soviets also began to become interested in the search for intelligent extraterrestrial life. In 1966, Sagan and Soviet Shklovskii published *Intelligent Life in the Universe*, as we observed in Chap. 3, which, in addition to dealing with life in our solar system and outside of it, dealt with possible radio, optical and even direct contacts between galactic civilizations. The Ozma project was followed by several others. The *SERENDIP* (Search for Extraterrestrial Radio Emissions from Nearby Developed Intelligent Populations) project started in 1979 and was followed in 1985 by the *META* (Megachannel Extra-Terrestrial Array) *project*, whose spectrum analyzer had a capacity of 8 million channels. The US government also entered SETI programs and financed NASA's *MOP* (Microwave Observing Program). The purpose of the program was to perform a targeted search for 800 specific nearby stars. The program was short-lived; it was in fact canceled the following year, but the project restarted with the name of the *Phoenix Project*, which was supported by private funding sources; furthermore, the number of stars observed increased to 1000. Even the *BETA* (Billion-Channel Extraterrestrial Array) *project, which is* 1000 billion times more powerful than the instrumentation used in the Ozma project, was short-lived, whereas the *ATA* (Allen Telescope Array) *project* was more successful. It began operation in 2007 with an array of specialized radio telescopes for SETI studies consisting of 42 antennas and was expanded at the end of 2010, reaching 350 antennas arranged over an area of 1 km in diameter, functioning as a single radio telescope. The project searches for signals in the 20,000 closest red dwarfs that are candidates for hosting the most ancient civilizations. In 1999, UC Berkeley began another project called *SETI@home*. With this project anyone can be involved in SETI research, simply by downloading software from the internet. The project uses observation data from the Arecibo radio telescope. The data are stored and sent to the SETI@home servers. The data are subsequently divided into small pieces and analyzed via software to identify signals. Each block of data is analyzed by the volunteers' computers, which then send back the result of the analysis. In this way, what appears to be a very onerous problem in terms of data analysis is reduced to a much more reasonable problem owing to the help of a large community of volunteers. In

2004, SETI@home II was released, but after more than twenty years, in 2020, the project announced its closure, at least in its current form. A contribution to SETI also comes from the Chinese radio telescope *FAST* (Five Hundred Meter Aperture Spherical Telescope), which since 2016 has been the largest in the world, and in a few years, the *SKA* (Square Kilometer Array) radio telescope should become operational, in which approximately one hundred organizations from twenty countries collaborate. SKA consists of thousands of antennas located in Australia and South Africa and has a surface area of one million square meters. As already mentioned, it will be able to detect the radar of an airport on a planet located 50 years away. As mentioned in the introduction, the *Breakthrough Initiatives*, a 10-year project, was founded in 2015 with the aim of searching for extraterrestrial intelligence. One of the program's projects is *Breakthrough Listen*. Three telescopes will be used, the Green Bank one, the Parkes observatory, and the Lick observatory optical telescope, which will study a million stars of the Milky Way and a hundred nearby galaxies.

11.2 OSETI Projects

While most SETI experiments observe the sky in the radio wave spectrum, some researchers have considered the possibility that alien civilizations may use optical emissions. For communication to be possible with optical radiation, it must be very intense. This can be achieved with laser technology, generating short pulses of light powerful enough to be recognized at great distances. The idea was presented for the first time in the scientific journal Nature in 1961 and in detail in the *US journal Proceedings of the National Academy of Sciences* by Charles Townes, one of the inventors of the laser. This type of SETI carried out in the optical field is called *OSETI*, i.e., optical SETI. One of the peculiarities of OSETI is that the pulses to be searched for are much faster and have a wider bandwidth than the radio pulses. To capture these signals, medium-sized optical telescopes are sufficient within the reach of any amateur. Searching for signals at optical frequencies presents two problems. The first problem is that while radio waves can be emitted in all directions, lasers are highly directional. This means that a laser beam could be blocked by a cloud of interstellar gas, and it could be observed only by terrestrial observers if it pointed toward them. The other problem is that lasers emit light of only one specific frequency, making it difficult to imagine which one you should listen to. However, this problem can be solved with mathematical techniques.

In the 1980s, two Soviet researchers conducted a short OSETI search, which produced no results. During most of the 1990s, OSETI research was limited by the observations of Stuart Kingsley. Numerous OSETI experiments are currently underway. A group of scholars from Harvard University and the Smithsonian Institution designed a laser detector and mounted it on Harvard's 155 cm optical telescope. Between October 1998 and November 1999, the research examined approximately 2500 stars. Nothing that looked like an intentional laser signal was detected, yet efforts continue. UC Berkeley is also conducting two different types of OSETI research. The first is directed by Geoffrey Marcy, the discoverer of exoplanets after Mayor and Queloz, and involves examining recordings of spectra collected during the hunt for exoplanets to look for laser signals that are continuous rather than pulsating. The second is more similar to what the group of Harvard universities and the Smithsonian Institute is aiming for and is directed by Dan Wertheimer of Berkeley. Between 2004 and 2016, a group of astronomers from the University of California at Berkeley, again as part of the SETI project, examined approximately 5600 stars in the Milky Way, of which at least 2000 are surrounded—or could be surrounded—by planets on which life would not be excluded. In addition, guess what the result is: nothing at all.

Thus, there is another problem. How can one understand if a signal is natural or produced by some extraterrestrial civilization? A first fundamental clue for a signal from a civilization is that it repeats itself. If you want to be identified, you do not send a single signal. We know that there are natural signals that repeat at regular intervals, such as the pulses of particular stars and pulsars. Therefore, the signal should have something that cannot be produced by a natural signal. For example, the signal should have, for example, mathematical or similar content. There is another way of checking whether a signal is natural or not, the so-called *Zipf's law* stated by George K. Zipf in a book by him in 1949. The law shows that for every word used with a certain frequency, there are approximately ten that appear one tenth of the time, one hundred that appear one hundredth of the time and so on. The law applies to all languages of the world and to the emissions of dolphins. Extraterrestrial languages are also thought to satisfy this need.

11.3 METI Projects

The SETI searches for signals coming from space, but we can also think of sending messages to extraterrestrial civilizations. This is an active SETI or *METI* (Messaging to Extra-Terrestrial Intelligence), an acronym coined in the 1980s by Alexander L. Zaitsev.

One problem for METI is the lack of a preestablished communication protocol. Typically, the messages created were based on symbolic logic expressing mathematical notions or using pictorial language. It was thought that counting, adding or subtracting should be a general ability. In 1960, mathematician Hans Freudenthal developed *Lincos* (Latin for cosmic language) as a possible language for use in radio transmissions to extraterrestrial civilizations. Lincos was used in 1999 and 2003 by the astrophysicists Yvan Dutil and Stephane Dumas to send messages to nearby stars. Later, other languages with similar purposes, such as *Astraglossa* and *CosmicOS*, were built. In addition to the communication language, METI has another drawback: the time required to exchange messages. If civilizations existed on the nearest star, Proxima Centauri, it would take 4.2 years for the message to arrive there and just as long for us to receive the response. A reasonable time. However, if civilizations were on the Andromeda Galaxy to send a message and receive a response, 5 million years would pass. METI, by sending radio messages, was unintentionally started with our radio and television broadcasts in the first half of the twentieth century. These signals were of modest strength and may have only reached a few nearby stars. The first intentionally emitted signals were sent in 1962 from Yevpatoria, on the Crimean Peninsula, directed toward Venus and bounced back toward Earth, and some of them are on their way to the star Gliese 581, which has three extrasolar planets.

In 1974, the so-called *Arecibo Message* was sent, a radio message transmitted into space from the Arecibo radio telescope in Puerto Rico and addressed to the Hercules Globular Cluster (M13), which is 25,000 light years away from us. Two simple reasons led to sending the message to M13: it is the brightest globular cluster in the Northern Hemisphere, and it was visible (even to the naked eye) in the sky at the moment in which it was decided to send the message; furthermore, it is part of a large, relatively stable constellation. The message, which was less than three minutes long and made up of 1679 binary digits, was created by Frank Drake and Carl Sagan and contained information on the atomic numbers of elements important for life, information on DNA, the population of the Earth, a graphic representation of a human body, a diagram of the Solar System with the position of our planet highlighted, etc. By the time the message reaches the globular cluster, 25,000 years from now, its nucleus will no longer be in its current position owing to its movement around the galactic center. However, the proper motion of M13 is so small that the message might still reach the center of the cluster. Messages were later sent to stars between 32 and 69 light years away.

To provide only a few examples, on July 6, 2003, signals were sent to several stars. The one directed toward the star 55 Cancri should arrive at its destination in May 2044, and the one directed toward 47 Uma should arrive in May 2049. In 2008, a signal was sent toward Gliese 581 c planet, on which conditions favorable to the development of life are supposed to exist. The signal should arrive on the planet in 2028, and if there is a response, we should receive it in 2048. In 2013, the *Lone signal project* was started, which allowed the continuous sending of short messages from individual people toward the red dwarf Gliese 526 at 17.6 light years.

In addition to the transmission of signals with radio telescopes, METI has carried out other projects. In 1972 and 1973, the Pioneer 10 and Pioneer 11 probes, respectively, were launched into the outer solar system.

Two 15 × 23 cm plaques depicting a human couple, our position in the solar system and other data were added (Fig. 11.1).

In 1977, Voyagers 1 and 2 probes were launched, and two gold plaques were also inserted (Fig. 11.2), which contained 115 images, 55 messages in different languages of the world, various sounds of Earth's nature and human sounds. Some, including physicist Stephen Hawking and physicist and

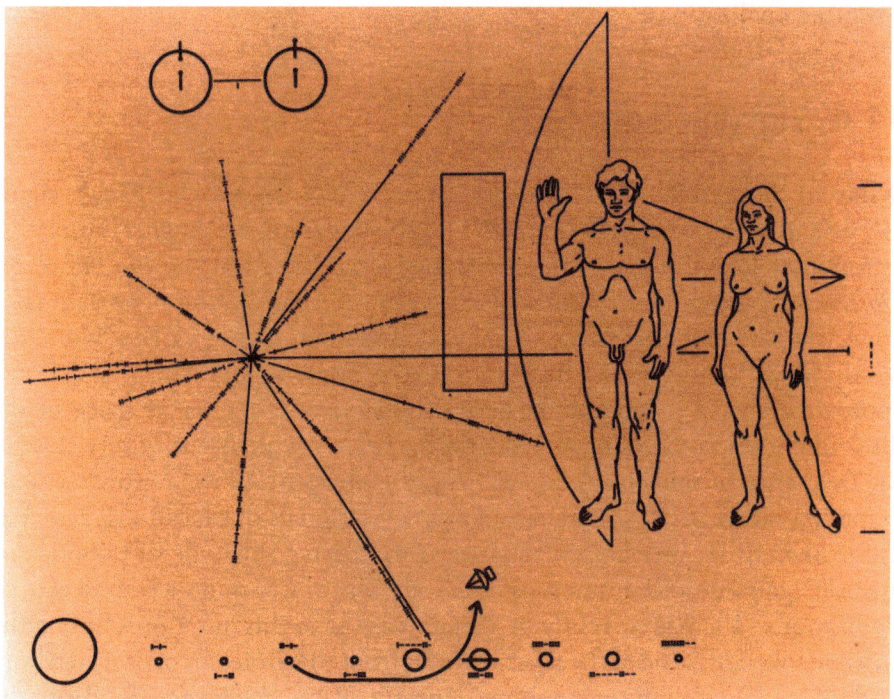

Fig. 11.1   Pioners' placque. (Credit: NASA)

Fig. 11.2 Voyager's disk. (Credit: NASA)

science fiction writer David Brin, have heavily criticized METI, as it risks endangering our planet by not knowing the level of development and intentions of extraterrestrials. Hawking highlighted that from our history, it can be deduced that the meeting of civilizations has brought great problems to less advanced civilizations.

11.4 Signals from SETI?

What are the results of SETI experiments? None, except the *Wow!*), signal 1977, and another signal from 2011. The morning of August 18, 1977, astronomer Jerry Ehman, while checking some data from the *Big Ear Radio Observatory*, discovered an unusual extraterrestrial signal. Out of great amazement, after having circled the signal with a red pen, the researcher wrote the word "Wow!". next to it, hence the name *Wow! signal*. The signal lasted approximately 72 s and should have originated in a region of space in the direction of the constellation of Sagittarius, with a peak intensity very close to the frequency of the 21 cm line of neutral hydrogen. All the researchers participating in the SETI program look for other signals similar to this one, but unfortunately, no one has ever managed to observe this type of radio transmission again, despite numerous attempts. Was this a sign of extraterrestrial civilizations? There is currently no certain answer to this question. Among the possible solutions, one of the most recent is from the professor of St. Petersburg College in Florida, the astronomer Antonio Paris. According to Paris, the signal may have been produced by two comets that, in 1977, were located very close to the source of the signal: 266P/Christensen and 335P/Gibbs. The signal would therefore have been produced by the hydrogen cloud that accompanies them and would no longer have been repeated in that same position because over time, the comets have slightly modified their orbit, no longer passing through that point of space again. However, this is just one of the many theories developed over these almost 50 years; the truth is that we still do not have a certain answer as to who or what produced that mysterious signal. Another signal on which there is no clarity is the one received between January and February 2011. The SETI reports the reception of 2 "unnatural" signals "of probable extraterrestrial origin", indicating that its antennas at 50 candidate planets were discovered a few months earlier from the Kepler mission. Since the signals were no longer repeated, it is assumed that they were due to terrestrial interference. However, SETI will continue to observe that region of the sky at other radio frequencies.

At this point, another question arises spontaneously: what could be due to this great silence? A possible answer could be similar to that given by Fermi in 1950, on which the Fermi paradox is based. There would simply be no extra-terrestrial life evolved to the point of being able to communicate or visit us. A simple calculation would prove Fermi right. We know that massive objects cannot reach the speed of light, which places a heavy threat on the possibility, even in the future, of traveling outside the solar system. The duration of the journey would require the departure of a community capable of self-sustaining and reproducing. Some estimates conclude that if the community could colonize the entire galaxy in 5–20 million years, it would be able to move at speeds equal to 10% or 1% of the speed of light, respectively, very short times on a cosmic scale. The absence of signal reception or a visit by alien civilizations to this planet would prove Fermi right. This would be in agreement with the *rare earth hypothesis*; that is, life on Earth is an exceptional phenomenon born from a myriad of random combinations. On this point of view, agrees the result of a paper of Snyder-Beatty and collaborators also in agreement with the original argument suggested by Brandon Carter that intelligent life in the Universe is exceptionally rare. However, the model, based on Bayesian statistics, is probably too simple to be significative. Bloetscher using Bayesian statistics conclude that the probability we are alone in the galaxy is significant, however the maximum number of contemporary civilizations might be as few as a thousand, which is not so small. The point of view that we are alone is however not accepted by much of the scientific community. Another possibility is the short duration of civilizations. We discussed this point in Chap. 10, discussing the term L, the average duration of technologically advanced civilizations. We have seen that there are many reasons why a civilization might disappear, both culturally and naturally. Total destruction due to global thermonuclear conflict or regression to primitive levels due to conflict. Impact from asteroids, comets, or supervolcanism can annihilate a civilization. A civilization may not even want to communicate even if it has developed the appropriate means, either because they are much more advanced than us and have no interest in communicating or because they may think that direct contact could harm both. Related to this possibility is the *dark forest theory*, according to which a civilization capable of space travel would regard all other intelligent life as an inevitable threat and would not attempt to communicate. Another possibility is the *theory of autarky* according to which civilization would thus have evolved perfectly autarchically and would therefore have no reason to expand. Another possibility is that extraterrestrials can communicate not via electromagnetic wave technology but via neutrinos or gravitational waves. However, if this civilization had evolved to this point, it should not be difficult to receive and decode radio signals, even if they are now obsolete. In 2018, Alexander Berezin of the National Research University of Electronic Technology put forward a new proposal, renamed "*First in, last out*", i.e., the first to enter is the last to leave. The study hypothesizes that civilizations that develop the ability to perform interstellar travel tend to eliminate others in their expansion attempts without necessarily having negative intentions. Another idea is that the existence of *the* great filter is an obstacle that prevents contact with extraterrestrial civilizations and their evolution. To understand what this is about, we need to examine the *Kardashev scale*, a well-known classification of intelligent civilizations. Those of Type I have the ability to exploit all the energies available on the planet that hosts them. According to Carl Sagan, we would have arrived at Type 0.7. Type II vehicles can exploit all the energy of the star at the center of their system. Type III algorithms can exploit the entire galaxy, which is inconceivable for our current status. At a certain point in the development of a civilization, before reaching Type III, there may be a kind of barrier against which all civilizations virtually collide: a phase, in the course of the long evolutionary process, that is impossible or at least very difficult to overcome. The Great Filter would be a sort of obstacle that inhibits the development of long-lasting extraterrestrial civilizations. Another possibility is that civilizations are too distant in space and time. Leaving aside the idea of entire communities organizing to conquer the galaxy, receiving signals from a high civilization in our own galaxy is prohibitive. If we want to be optimistic and assume that there are 1000 advanced civilizations in our galaxy, we can estimate the average distance between each civilization to be approximately 1992 light years. This makes it very difficult to communicate with these civilizations. By sending a signal, we receive a response after 3984 years. If the number of civilizations was 10, the average distance between them would be 9244 light years, and so on, and sending a simple "Hello" and receiving a response would take 18,488 years. In summary, even if we had many civilizations in our galaxy, we would not have the ability to contact them.

11.5 Peculiar SETI Methods

The SETI, OSETI, and METI searches did not provide us with particularly interesting results. Studies of extrasolar planets are in their infancy but promising. In the coming decades, we could scan the atmospheres of those planets in search of biomarkers. Moreover, scientists interested in the search for life in the Universe are looking for other ways to detect alien traces. With the METI projects, we have launched some of our probes into space with messages. There are those who have thought that extraterrestrials may have done something like this. On October 18, 2017, Rob Weryk, a member of the team working on the *Pan-STARRS* (Panoramic Survey Telescope & Rapid Response System), an exploratory system of celestial bodies developed and managed by the University of Hawaii, discovered the first object coming from another star system called: *1I'/Oumuamua*. The number 1 indicates that it is the first cataloged object of this type, the I comes from the indication Interstellar, whereas 'Oumuamua means "messenger who arrives first from afar" in the Hawaiian language. Its unusual trajectory and its relative speed with respect to the solar system of approximately 26 km/s immediately made it clear that it was an interstellar object probably coming from approximately the direction of the star Vega. Its shape was atypical; it looked more like a torpedo, several tens of meters long, than an asteroid. The amount of light coming from it varied by a factor of 10 every 5 h. Some observations have led to the hypothesis that on the surface of 'Oumuamua, there is a layer of organic material approximately 50 cm thick, which prevents the sublimation of the ice contained within it, thus hindering the formation of the wake that would be expected with this type of trajectory with respect to the Sun. During the passage to the minimum distance from the Sun, a slight nongravitational acceleration was recorded. A possible explanation for this acceleration is that it is due to modest cometary activity. It appears that the object has developed a modest coma produced by the melting of ice on it, caused by solar radiation. Another explanation is that the acceleration was generated as an effect of solar radiation if 'Oumuamua was between 0.3 and 0.9 mm thick, i.e., it was an artificial "solar sail". In a paper published in 2018, Abraham Loeb and Shmuel Bialy argued that to account for 'Oumuamua's gradual acceleration, the object had to be less than a millimeter thick and at least twenty meters in diameter. Loeb claims that 'Oumuamua could be an alien artifact. This hypothesis remains essentially speculative. A recent study conducted by two astrophysicists from Arizona State University, Steven Desch and Alan Jackson of the School of Earth and Space Exploration, established that it was a fragment of a Pluto-like planet from another planetary system.

Extraterrestrial intelligence can be detected via radio telescopes, but it can also be inferred from anomalies near a planet, indicative of what astrophysicist Nikolai Kardashev called type II civilizations, which are capable of exploiting all the energy available from its parent star. To exploit this energy, a particular structure called a *Dyson sphere* (Fig. 11.3) could be built. In 1960, Freeman Dyson published an article in the journal *Science* entitled *Search for Artificial Stellar Sources of Infrared Radiation*. In the article, Dyson theorized that technologically advanced societies could completely surround their home star to maximize the capture of energy coming from the star. This spherical structure intercepts all visible light and sends it inward, whereas all unused radiation is sent outward in the form of infrared radiation.

From this, it follows that a possible method to search for extraterrestrial civilizations could be the search for large sources of infrared emission. In 2015, researchers at Penn University took up this idea again and studied more than 100,000 stars with the help of NASA's *Wise infrared telescope*. In fact, if

Fig. 11.3 Artistic image of a Dyson's sphere

an extraterrestrial civilization had colonized a galaxy, it would use so much energy that infrared emissions would be detectable even by us. A coauthor of the research, Roger Griffith, explained:

Our work began by searching for galaxies in more than 100 million objects detected by Wise, within which we identified approximately 100,000 candidates. Among these, approximately 50 showed strong mid-infrared emission.

Unfortunately, subsequent analyses revealed a natural cause for these emissions. Griffith continuing said

The result is interesting because many of the galaxies studied are billions of years old, and therefore, within them, there would have been plenty of time for very advanced civilizations to develop. Our conclusions are that advanced civilizations do not exist, or if they do exist, they are not advanced enough to emit significant amounts of energy.

In short, according to these researchers, if extraterrestrial civilizations exist, they live without consuming energy in detectable quantities. Another similar study, with more positive results, has already been carried out by researchers at Fermilab in Chicago using the IRAS satellite to discover anomalies in the infrared part of the spectrum. Out of 250,000 stars, 17 candidates have been found, and among these, 4 have variations that are difficult to explain through natural phenomena. The best candidate is IRAS 20369--5131. The promoters of the research claimed that with IRAS data, it could be possible to discover Dyson spheres up to 1000 light years away. Hongying Chen, a postdoc from the Chinese NAO Institute and M.A. Garret, studied the infrared emission of 21 galaxies. While the origin of the emission for nineteen of them is natural, for the remaining two, ILT J134649.72+542621.7 and ILT J145757.90+565323.8, the situation is unclear. In this study, the two researchers concluded that they could be inhabited by Type III civilizations. There are similar speculations related to the Bootes void. In the Universe, there are more or less spherical regions that contain very few galaxies. The Bootes Void is located 700 million light years away, has a diameter of over 330 million light years and contains only 60 galaxies. It is so large that if the Milky Way had been right at the center of the void, it would have taken humans until the 1960s to discover the existence of other galaxies. According to modern estimates of the age of the Universe, the largest void should be on the order of tens of millions of light years in diameter. The Bootes void is approximately 10 times larger than it should be. Its origin can be explained in a natural way through the coalescence of smaller voids. However, some speculations state that when we look at the Bootes Void, we see nothing because enormous Dyson spheres or similar structures have been built on thousands of galaxies, preventing the starlight from reaching us and making the region seem empty. In other words, the Bootes Void would be made up of a series of galaxies inhabited by a Type III civilization. However, by making simple calculations that consider the dimensions of the void, the species that would occupy it would not have time to create the offspring necessary to colonize all that space.

Some glimpsed anomalies indicative of a Type II civilization in the strange fluctuations of light that were received starting in October 2015 from the star KIC 8462852 or Tabby's star, in honor of the astronomer Tabetha Suzanne Boyajian. The Kepler space telescope showed dips in the star of up to 20% at nonperiodic intervals, which could not be associated with the presence of a stellar companion or any planet.

The irregular variations in the star's brightness are compatible with a large mass or a collection of many small masses orbiting the star. Some hypotheses have been proposed to explain a star's unusual emission profile, but none are universally accepted. In October 2015, Jason Wright proposed the hypothesis that the unusual variation in light emission could be associated with intelligent extraterrestrial vision. According to Wright, the objects that eclipse the star may belong to a megastructure built by an alien civilization, such as a Dyson sphere. The SETI Institute began on October 19, 2015, to point the satellite dishes of the ATA (Allen Telescope Array) toward the star to search for radio emissions coming from intelligent extraterrestrial life without success. In October 2017, after long studies, NASA announced that the variations in the star's brightness are due to a disk of dust and other materials with a very irregular and mobile structure. Subsequent studies confirmed this hypothesis. From the analyses, the researchers noted that whatever material exists between us and Tabby's star would block more blue light than red light. The only explanation, therefore, that remains standing is space dust. However, even in space dust, there is something that does not seem to add up. If it was a ring of dust around the star, it would constantly block the starlight rather than generate brightness fluctuations. The amount of dust needed would have to be greater than Tabby's star would be able to produce. Boyajian said of the star, "*We're certainly not done with this star yet.*" A phenomenon similar to that of Tabby's star has been observed around the red dwarf star EPIC 204376071. The celestial body was subject to a surprising dip in brightness, which closely resembled Tabby's star. Like the latter, this strange star may be obscured by an orbiting cloud of dust, a remnant of the protoplanetary disk from which planets are born around stars. This is, however, only one of the possible hypotheses, which is extremely difficult to confirm with the small amount of data available. Another possible explanation could be the presence of a large gaseous planet surrounded by rings, considerably larger than those of Saturn, but this last theory could soon be abandoned given the little coincidence between the computer simulations and the observations made. In short, what is positioned between the star and our observation point essentially remains a mystery. Maybe a Dyson Sphere?

For millennia, man has had to limit himself to speculation on the existence of extraterrestrial life and intelligence. In the twentieth century, the study of the possibility that life existed outside the Earth was discredited, even by esteemed scientists such as Fermi. Today, things are changing rapidly, and the discovery of extrasolar planets has contributed to this. With the most powerful new exploration projects ever undertaken, in a few decades, if extraterrestrial life existed, we would be able to detect it. We have already shown that Epicurus was right when he wrote

> *Furthermore, there are infinite worlds both like and unlike this world of ours. Because the number of atoms is infinite, as has already been proven, they are borne on far out into space. For those atoms, which are of such a nature that a world could be created out of them or made by them, have not been used up either on one world or on a limited number of worlds, nor again on all the worlds which are alike, or on those which are different from these. Thus, there nowhere exists an obstacle to the infinite number of worlds.*

and perhaps we will be able to prove that Giordano Bruno was right when he wrote

> *Thus, there is not merely one world, one earth, one sun, but as many worlds as we see bright lights around us …. In which other inhabitants move, live, vegetate and put into effect the acts of their vicissitudes.*

What we discover will not only change our vision of our being in the Universe but also the way we view ourselves.

12

We Are Not Alone

Despite decades of studies, we have no evidence of the existence of life in the solar system or on extrasolar planets. This could change in the next decade with observations from the James Webb Telescope or the *ARIEL* (Atmospheric Remote-Sensing Infrared Exoplanet Large-survey) *mission*, the project of a space telescope for the study of extrasolar planets, with their physical conditions and chemical compositions, which is expected to be launched in 2029. It will be very unlikely that anything new will come from the SETI, OSETI, and METI projects for all the reasons discussed in the previous chapter. Therefore, the following question remains: are we alone still standing? Obviously, there are studies that have given a probabilistic answer to the question by trying to overcome the limits on knowledge of the parameters of the Drake equation. Returning to the Drake equation, we have three astrophysical parameters that are quite well measured today (R^* f_p n_e). The factors f_l, f_i, and f_c that have to do with the emergence of life, intelligence, and technology are not known, as is the duration of civilization, the L factor. In 2016, Adam Frank and Woodruff Sullivan published a paper in the journal *Astrobiology*, reviewing the Drake equation in light of Kepler's discoveries. The two scientists reformulated the starting question in such a way as not being interested in the average duration of a civilization, the L factor, or whether this civilization still exists to be able to receive the possible messages sent by it. Their choice distances their study from the formulation of the Drake equation, which aims to calculate the number of existing technological species. The question we ask ourselves is therefore: *what are the chances that ours is the only technologically advanced civilization that has ever existed?* This change in perspective reduces the uncertainty terms present in the Drake equation. As already mentioned, the 3 astrophysical parameters are well estimated, and the three terms f_l, f_i, and f_c remain in the equation. By eliminating the L factor and taking into account the entire Universe, not just our galaxy, the Drake equation transforms into the product of the factors $N = (N^* \; f_p \; n_e) \; (f_l \; f_i \; f_c) = N_{astrophysics} \; f_{bt}$. In other words, the number of civilizations that existed in any epoch in the Universe are given by the product of astrophysical factors, $N_{Astrophysics}$, with N^* the total number of stars in the Universe, which is on the order of 2×10^{22}, f_p approximately 1 and n_e approximately 0.2, as in Chap. 10, and from f_{bt}, which collects the factors related to the birth of life, intelligence and technology. The two scientists worked in statistical terms and estimated the lower limit of the probability that one or more technological civilizations evolved at some place and time in the observable Universe. In terms of probability, if N was equal to 0.01, this would mean that if the history of the Universe was repeated 100 times, only one technological civilization would appear. The important result they obtained is that unless the probability that an alien civilization has developed on a habitable planet is less than one in a million billion billion, 10^{-24} (i.e., 0.000000000000000000000001), which is a truly small number, humans are not the first technologically advanced life form to have inhabited the observable Universe. Repeating the calculations for a galaxy such as ours, we obtain that the probability is equal to 1.7×10^{-11}; that is, we are sure that a technological species has developed in the history of our galaxy if the probability that a technological species appears on a habitable planet is greater than 1 in 60 billion. In the past, when the Drake equation was used, more or less pessimistic hypotheses have been formulated about the formation of civilizations on other planets. One of the most negative claims is that the probability of a civilization forming is 1 in 10 billion for each planet. Taking into account this pessimistic estimate and Frank and Woodruff's result for the entire Universe, trillions of technological civilizations would have existed in the history of the Universe. Obviously, the study does not refer only to the past but is valid for future eras; therefore, we must expect that civilizations have existed before us and will exist after us. Before Frank and Woodruff, Amir D. Aczel, a few years after the discovery of the first exoplanet, published a book, *Probability 1*, in which he calculated the probability of existence of a planet with life in the Universe via statistics. Unlike Frank and Woodruff, Aczel wanted to answer the following question: are we alone? not if civilizations existed throughout the whole history of the Universe. He assumed $f_p = 0.5$ in the Drake equation; today, we know that it is approximately 1, and he assumed that among the exoplanets known at the time (only 9), there was at least one in the habitable zone. He assumed that the probability of life originating is very low: 1 in 10^{12}, that the number of stars in our galaxy is 300 billion, that there are 100 billion galaxies, and that the number of stars in the universe is on the order of 3×10^{22} and used an elementary statistical rule, that of the union of independent elements, to find the probability of life around a star in the Universe, finding that the probability is $P = 1-(0.999999999999995)^{30\,000\,000\,000\,000\,000\,000\,000\,000}$, i.e., a probability indistinguishable from 1, i.e., 100%. The same result would be achieved even if there were 10 billion stars in our galaxy and if a billion galaxies existed. Even though the probability on a single planet is very low, the compound probability that life exists on at least one planet increases steadily due to the large number of stars and planets. Another thing to add is that today, we know that our universe has a flat geometry; therefore, it could be infinite, and this further increases the probability. In 2020, another statistical study by Amedeo Balbi and Claudio Grimaldi evaluated the impact of a discovery of life on a single planet on the number of planets on which there is life. According to the study, if a planet with life was found, there should be at least 100,000 in our galaxy. A few days ago (February 14, 2025), a groundbreaking study suggested that life may be a common outcome on planets, rather than a rare occurrence. As one of the co-authors, Jason Wright, says "This new perspective suggests that the emergence of intelligent life might not be such a long shot after all. Instead of a series of improbable events, evolution may be more of a predictable process, unfolding as global conditions allow". To confirm these studies directly, we must wait for the next few decades to study the atmospheres of habitable extrasolar planets. These studies could also provide more precise numbers via the Seager equation, for example. One way to overcome the parameter problems of the Drake equation is the equation introduced by Sara Seager. If one needs to know the number of planets with detectable signs of life, one needs to know the number of stars observed, the number of stable stars, the fraction of stars with rocky planets in the habitable zone, the fraction of those planets that can be observed, the fraction with life present, and the fraction on which life produces gaseous biosignatures detectable. The equation focuses on finding planets with biosignature gases, which are gases produced by life that can accumulate in a planet's atmosphere to levels that can be detected with remote space telescopes rather than aliens equipped with radio technology. In other words, this equation, with the observation of powerful space telescopes, makes it possible to reveal whether life exists on a planet. In summary, compared with the times in which Drake began his studies, today, we have much more information on the astrophysical parameters that serve to answer the question of whether there is or has been life in the Universe. Using this new knowledge and statistics, we can answer our question, and it seems that there have been civilizations in the history of the Universe. We probably will never have answers to our question using methods like SETI, but in a decade or two with the study of the atmospheres of nearby habitable exoplanets, we will be able to say with certainty whether there is life on them. Therefore, the answer to the question is as follows: have there been and will there be civilizations in our Universe? is almost certainly positive. However, it is very likely that we will not have direct contacts, as in the film *Contact* or *Close Encounters of the Third Kind*, or indirect contacts with exchanges of electromagnetic signals.

13

Epilogue

Several decades ago, when I looked at the sky and dreamed of being able to move through the Universe, I was guided by imagination and saw a future similar to what films such as Star Trek show: interstellar or intergalactic travel at speeds much greater than the speed of light. In recent decades, this has not been realized, but from a scientific point of view, many changes have occurred. We know much more about our Universe, about the planets that populate it, which at the time we did not even know if they existed. We found planets around stars similar to our sun but mostly planets around red dwarfs, which are by far the most numerous stars. The approximately 5000 planets discovered are only a handful of objects, the discovery of which made us understand that planets are not rare objects in space but rather common objects and that every star has planets. The universe is full of planets. In our galaxy alone, there are hundreds of billions, and as we have seen, an estimated 6 billion are similar to Earth, which revolves around stars similar to the Sun. If we move on to red dwarfs, half of this population of stars should have a habitable terrestrial planet and an average of one habitable terrestrial planet every 5–6 light years. It is no coincidence that the closest star to us, Proxima Centauri, has a habitable planet, Proxima Centauri b. With the billions of terrestrial-type planets that exist, for statistical reasons, many of them must be in the habitable zone. What about life in the Universe? If things go as in the case of Earth, on which life appeared a few hundred million years after its formation, it must be present on many habitable planets in our galaxy and in the Universe. It probably will not be the life we would like to find, the strange humanoids of Star Wars or similar films; it will be bacteria or who knows what else, but it is still life. The most interesting thing is that if we had visited our planet a few billion years ago, we would not have found cows, horses, or men but rather simple bacteria and that simple life has evolved over billions of years to give rise to increasingly complex and more intelligent beings. Why could this not happen on other habitable planets, especially those that revolve around orange dwarfs that can have a lifetime of 20–40 billion years instead of the 10 billion of our Sun? As we have seen, starting from statistical studies based on knowledge of the astrophysical parameters of extrasolar planets, we conclude that in the life of the Universe, technological civilizations have existed until today. From the knowledge we have gained, it is easier to say that we are not alone than the opposite. This almost certainly clashes with the great silence that has been given to us by decades of research with electromagnetic waves of signals from advanced civilizations. Fermi's idea inherent in his paradox that there are no intelligent extraterrestrial life forms can no longer be accepted as lightly as in the 1950s and, similarly, the *Rare Earth hypothesis*. The nonreception of signals from extraterrestrial civilizations can be due to many reasons, first, the enormity of space, the lack of knowledge of the place where these civilizations may be located, or the duration of these civilizations. On the other hand, if someone had tried to contact us by sending us electromagnetic signals in the last approximately 4.5 billion years, they would certainly not have found us. We have only been sending and receiving signals for a hundred years. Therefore, in addition to the enormous space between the stars, which makes contact difficult, we must add a time factor. Extraterrestrial civilizations may not be interested in communicating, or they may use other technologies. Another thing to add is that we talk about life but not only have we not been able to find the mechanisms by which it was born on Earth, but we also have difficulty defining it. Even if we do not have a good definition of life, most likely in the next decade, the study of the atmospheres of habitable planets with the space telescopes we have, and those we will have, will allow us to identify planets in which life releases gases that testify to its existence. Primitive life, such as that which existed on Earth long ago, changed the Earth and transformed it into Gaia, a sort of superorganism capable of self-regulating and generating conditions suitable for the development of life, owing to the activity of living things themselves. This discovery, however, would also lead us to a real paradigm shift and would make us understand that we are not unique in the Universe in this context either. As Arthur C. Clarke noted, there are two possibilities: *"We are alone in the Universe or we are not. Both are shocking"*. However, I am convinced that if life managed to make its way onto a planet such as Earth, to become extinct and reborn, to evolve and achieve self-awareness, there is no reason why this has not happened, is not happening and will not happen in other places in the Universe.

Appendices

Appendix A: DNA and Protein Synthesis

The structure of DNA was discovered by Watson and Crick: it is a double helix. Each strand is made up of repeating units, called *nucleotides*, made up of a sugar (*deoxyribose*), one or more phosphate groups and a nitrogenous base. There are four *nitrogenous bases*. Two of the nitrogenous bases are hexagonal rings of carbon and nitrogen to which hydrogen atoms bond. They are called *pyrimidines* and are thymine (T) and cytosine (C). The other two substances are made up of two rings of carbon and nitrogen welded together. These substances are called *purines* and are called adenine (A) and guanine (G). As already mentioned, these four molecules, A, G, T and C, are also known as nitrogenous bases. The individual's genetic code is written in DNA by a combination of these four molecules. The problem is understanding how these molecules are linked to the structure of DNA and which combination constitutes language. Molecules A, G, T and C are connected in pairs to each other and to the DNA strands. The four elements on which genetic information is based are precisely the nitrogenous bases, the molecules A, G, T and C. These connect to each other in a precise manner. The adenine, A, connected to one of the two strands binds only to the thymine, T, connected to the other strand. The two base pairs are joined by two hydrogen bonds. The same occurs for cytosine (C) and guanine (G). The complete structure of DNA is similar to a spiral staircase in which the uprights are the DNA strands and the rungs are the nitrogenous base pairs. The bases are separated by a

A. Del Popolo, *Extraterrestrial Life*, https://doi.org/10.1007/978-3-031-83497-4

distance of 3.3 angstroms (1 angstrom equals 0.00000001 cm), and the double helix is 20 angstroms in diameter. All the information of a living being is contained in this microscopic structure. The bases are matched according to a precise rule. In the case of DNA, as already mentioned, A is paired with T and C with G. Thus, if one strand of the double helix begins with AGGTCCGTAATG…, the other will be TCCAGGCATTAC. That is, by knowing one strand, one can deduce the other. The sequence of bases carries a message with this four-letter alphabet that conveys information for a protein. An amino acid is encoded by a group of three nucleotides, which is called a *codon. Genes* are sequences of nucleotides that contain complete information for a certain property. The human genome[1] is estimated to be composed of approximately 50,000 genes, is 3.5 billion "letters" long and contains 3 billion bits of information. Genes are contained in *chromosomes,*[2] which are contained in the cell nucleus. In DNA, the sequence of bases is encoded in sequences of three-letter words (for example, GGT, AGA, etc.). Since there are four nucleotides, 4^3 possible triplets are available to code for the 20 amino acids used to build proteins. Genes largely determine how we are made: our height, our weight, our appearance, our intelligence, etc. After the discovery of DNA, the problem arose of how it transmits its characteristics from one generation to another, i.e., how it replicates and how it generates proteins. The process is complex, and the details can be summarized as follows. Initially, the DNA stretches along its length, and the bonds that hold the two strands together begin to open like a zipper. At this point, another DNA-like nucleic acid, RNA, comes into play. The latter is made up of only one strand, not two like DNA; furthermore, RNA has a sugar *ribose* instead of the deoxyribose of DNA, and instead of the T base of DNA, it has the base *uracil* (U). The information from DNA is transcribed onto a strand of RNA, *messenger RNA* (mRNA). It passes through the cell nucleus and moves into the cytoplasm. Inside the cell, there are organelles called *ribosomes* (Fig. A1.1), *which are* also made up of another RNA, called *ribosomal RNA* (rRNA). As previously mentioned, messenger RNA (mRNA) contains information that is translated in the ribosome. The mRNA strand can be thought of as magnetic tape, whereas the ribosome is a type of machine that builds a protein from the information on the tape (mRNA). To this end, the ribosome moves along the mRNA

---

[1] The genome can be compared to the software of a computer and the individual genes to the instructions for making the machine (the organism) work, but which are also used to build it. In other words, the genome is a sort of instruction manual that first directs the development of our organism (embryo-fetus-newborn) and then the functioning of the organism itself.

[2] During the reproductive process of the cell, each unit of DNA, after duplicating itself, compacts into a structure called a *chromosome* and is passed on to daughter cells.

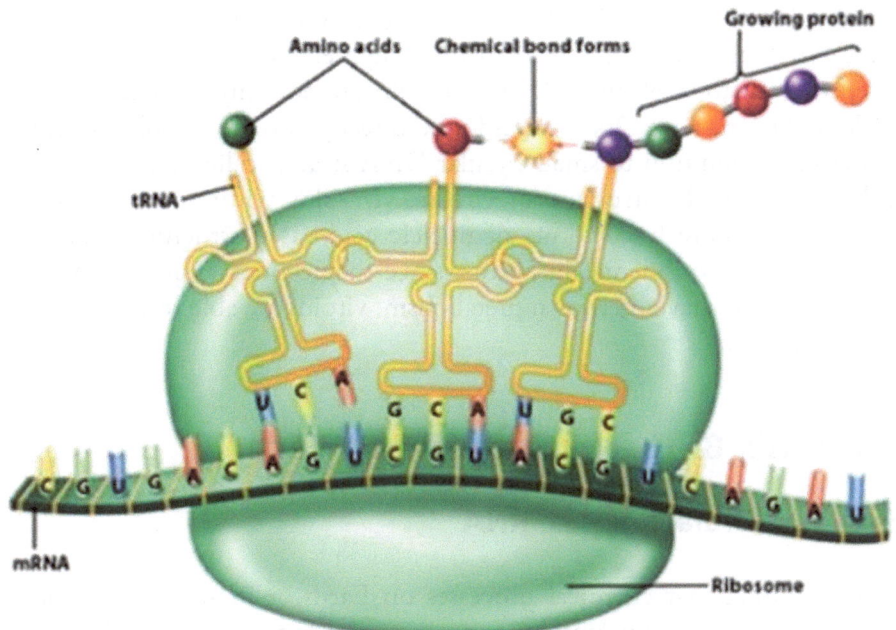

Fig. A1.1 Ribosome and translation of messenger RNA into proteins

filament and "reads" the information contained on the ribbon (mRNA) as it passes.

In this step, the codons are read in the order in which they appear on the mRNA. The ribosome then finds the amino acids corresponding to the codons, amino acids that are located near the ribosome. They are attached by particular bonds to another RNA, called *transport RNA* (tRNA[3]), which are molecules that have the shape of crosses. After the ribosome reads a particular codon, it searches for the corresponding anticodon,[4] hooks it and removes the amino acid that is attached to it. The ribosome combines this amino acid with the others already put together, giving rise to the protein. In short, a new strand is synthesized along one of the DNA strands, and the same happens to the other strand. The two new strands, copied from the original DNA, combine to form a new DNA molecule. In the construction of this new DNA, the rules relating to nitrogenous bases are as follows: A attaches to T and vice

---

[3] Transport RNA (tRNA) connects and relates the information contained in 3 nucleotides (called codons) of the mRNA with the amino acids of the proteins. The tRNA reads the codons and provides the amino acids corresponding to those letters.

[4] The anticodon is the triplet of *bases* present in the transport RNA (tRNA) with which the recognition of the triplet of the bases of the codon present in the messenger RNA (mRNA) occurs. If the codon consists of the AAC bases the anticodon is UUG. Remember that G pairs only with C and A only with U.

versa, and G attaches to C and vice versa. In this way, the structure of the parent DNA molecule is preserved in the daughter molecules. For precision's sake, we want to recall that DNA and RNA are not only present in the cell nucleus. Bacteria have DNA that is found directly into the cytoplasm without a nucleus, in addition to small circular DNA strands called plasmids. DNA and RNA form the structure of viruses, virusoids (viruses of viruses) and viroids, even more basic forms of nucleic acid-based structures. Although viruses are not considered life forms, since they cannot reproduce on their own and do not have their own metabolism, viroids are able to reproduce on their own.

Appendix B: The Birth of Life on Earth

From Primordial Soup to RNA

Even assuming that chemical processes on Earth or in space can generate amino acids, sugars, nucleotides, etc., there is a long way to go from these substances to the formation of life and the so-called LUCA. For the functioning of the ancestral organism, metabolism, i.e., the ability to extract energy from the surrounding medium and use it to stay alive, is needed. Metabolism is based on proteins that are fundamental to life processes, i.e., *enzymes*. An enzyme is a protein that can catalyze (i.e., accelerate) the speed of biological reactions in an organism. These enzymes are synthesized from the information present in nucleic acids. Unfortunately, for nucleic acids to duplicate and express information, enzymes are needed. We find ourselves faced with the chicken and egg problem in the field of the origin of life: enzymes are produced by nucleic acids, but for the latter to duplicate, enzymes are needed. It is necessary for the components of life to form all together and collaborate to generate it. How can this problem be solved?

Leslie Orgel suggested how to simplify the problem and suggested that primordial life had no proteins or DNA. The engine of life is composed almost entirely of RNA. For this purpose, the primordial RNA molecules had to be very versatile, and first, they had to be able to make copies of themselves. The idea that life began from RNA proved to be very influential. RNA can do something that DNA, a rigid double-stranded structure (double helix), cannot do. As a single-stranded molecule, RNA can fold into a variety of shapes, and such folds seem similar to the way proteins that are long strands of amino acids, rather than nucleotides, behave. A suspicion arose in Orgel's mind. If

RNA could fold like a protein, perhaps it could form enzymes. If this were true, RNA would be capable of storing information and, at the same time, catalyzing reactions, such as enzymes do.

In 1982, Thomas Cech and, in 1983, Sidney Altman showed that these ideas made sense, showing that some RNAs have catalytic capabilities (like enzymes) and that they can function as enzymes. Currently, the idea that life began with RNA seems promising. These RNAs with enzymatic properties are called *ribozymes*. In other words, a ribozyme is an RNA molecule capable of accelerating a chemical reaction, similar to enzymes. This discovery has important implications for our discussion of the formation of life. In fact, it is no longer necessary that the proteins and the nucleic acids that encode them are formed at the same time. This leads to the idea of the so-called *RNA world*, a term coined by Walter Gilbert. According to Gilbert, the first stage of evolution consists of "*molecules of RNAs that carry out the catalytic activities necessary to assemble themselves from a soup of nucleotides*".

In such a world, RNAs are capable of doing the things important to the formation of life, namely, carrying genetic information and functioning as catalysts for chemical reactions.

The RNA world is an elegant way of reproducing the complexity of life from scratch. Instead of relying on the simultaneous formation of large numbers of biological molecules from the primordial soup, some sorts of "jack-of-all-trades molecules" could do the job of all of them. The RNA world is accepted today because evidence has been found in favor of it. Importantly, as we observed in Chap. 2, the key reaction in the production of proteins, which is based on the union of the amino acids that constitute them, is catalyzed by an RNA called ribosomal RNA (rRNA). In 2000, Thomas Steitz's team produced a detailed image of the structure of the ribosome, which revealed that RNA is the catalytic core of that molecule. Thus, the discovery of the enzymatic capabilities of RNAs implies that the RNA world involves complex metabolism. The organisms of this world would undergo evolution through natural selection. Although the existence of organisms capable of evolving has been confirmed, some open points remain, such as the problem of the invention of the genetic code and the synthesis of proteins. Furthermore, the experiments are a long way from producing RNA. Remains the problem of how it could have formed on the primordial Earth. Gerald F. Joyce and Leslie Orgel argued that the spontaneous appearance of RNA chains on early Earth "*would have been something of a miracle*."

Before discussing this aspect, we would like to remind you that Orgel's idea and the RNA world were based on the assumption that a fundamental aspect of life is its ability to reproduce. The idea put reproduction at the forefront.

Indeed, other aspects, such as metabolism, are essential for life. For many biologists, the defining characteristic of life is metabolism, followed by reproduction. This seems pretty obvious, because before you can reproduce you have to be able to keep yourself alive. In the 1960s, scientists studying the origin of life split into two camps: those that polarized research on genetics and reproduction and those that put metabolism at the center. A third group argued that the first thing to appear was a container for key molecules, *cellularity*, or *compartmentalization*. Compartmentalization for these researchers came first. In other words, a cell must exist. The vital material must be enclosed by a membrane of fats and lipids. There are therefore three different research groups based on the different ideas enumerated: reproduction (the RNA world), metabolism, and cells.

Returning to the problem of RNA generation, the discovery of catalytic RNA led to the idea that it would account for at least two of the fundamental aspects of life: genetics and metabolism. The problem remains of explaining how RNA appeared. As we have already said, Orò showed that nucleotide bases (A, G, etc.) can be obtained in particular reactions, but the complex problem remains how to generate an RNA nucleotide (called a ribonucleotide) by connecting each base with the sugar (which, as mentioned in Chap. 2, is the ribose), which constitutes the RNA, with a phosphate group. Sugar (*ribose*) are formed in experiments, and the source of phosphate could be polyphosphates present in volcanic emissions or meteoric phosphates. Despite various attempts to obtain the necessary unions, it was not possible to carry them out. The other problem is that even if we manage to build nucleotides, we must then bind them to form chains or polymers. Furthermore, apart from the problem of RNA synthesis, its existence does not ensure the existence of an RNA world, since it must be able to replicate. There have been several attempts to replicate RNA in the absence of enzymes via two routes: without using enzymatic activity or with the use of RNA. In the first case, one starts from some RNA and nucleotides. The British chemist Leslie Orgel tried this for many years without success. Other researchers continued his work using chains of a few dozen nucleotides. Jack Szostak and Gerald Joyce are two of those who attempted the second path: using the enzymatic activities of RNAs. In 2014, there was a superreplicator that was very efficient in vitro but whose function is unlikely in the primordial Earth. Several researchers have continued to search for ways to replicate RNA. Despite all these attempts, the problem has not been resolved. The lack of self-replicating RNA was a fatal problem for the idea of the RNA world. RNA does not appear to be capable of giving rise to life. When it was taken for granted that it was impossible to get from the primordial soup to RNA, John Sutherland's research changed everything. Instead of trying to first form the sugar (ribose), the phosphate and the base and then join them, he used derivatives of cyanide and aldehydes in the presence of phosphate, obtaining cytosine (C) and uracil (U) nucleotides. Research has shown that hydrocyanic acid, together with hydrogen sulfide, ultraviolet light and copper ions, gives rise to precursors of nucleotides, lipids and 11 amino acids. These discoveries have brought us closer to the solution of the aforementioned chicken and egg problems, suggesting that, in some way, chickens and eggs could be formed and that the three essential characteristics of a minimal cell—information (genetic), metabolism and cellularity—could be achieved simultaneously.

Starting from Metabolism

The RNA world theory is based on the idea that the most important thing for an organism is to reproduce. However, many researchers do not believe that reproduction is fundamental. Before reproducing, an organism must be self-sustaining. To stay alive, you need to absorb some form of energy. Many researchers believe that the starting point is metabolism. As already mentioned, we call metabolism the ability to extract energy from the surrounding medium and use it to stay alive. This process is so important that many researchers believe that it is the first thing life has ever done. There is therefore a research chain that is based on the idea that metabolism comes first. What do organisms that only have metabolism look like? One of the most interesting ideas is that of Günter Wächtershäuser from the late 1980s. For Wächtershäuser, the first organisms were completely different from everything we know and were not made of cells; they would have been acellular, and they had no enzymes, DNA or RNA. Both nucleic acid- and information-containing molecules were missing. However, they would have had a certain metabolism and a capacity for evolution. These compounds, similar to the terminal or intermediate products of metabolism, have a negative charge and are anions. Wächtershäuser imagined a stream of hot water rich in volcanic gases such as ammonia and traces of volcanic minerals flowing from a volcano. Chemical reactions began to occur where water flowed over the rocks. Metabolic cycles are created, i.e., processes in which one chemical substance is converted into another, until the initial substance is recreated. During this process, the system absorbs the energy used to start the cycle. These metabolic cycles do not resemble life. Wächtershäuser spoke of precursor organisms that could barely be called living. He developed his model in the 1980s and 1990s, in great detail, outlining which minerals were in play and the chemical cycles that took place. It was a theory that needed discovery to support its ideas. This discovery had previously been made in 1977 by a team led by Jack Corliss using a submarine near the Galapagos Islands. Corliss and collaborators observed volcanically active rock ridges rising from the sea. These ridges were covered with hot springs. These areas are populated by a multitude of different types of animals. In other words, the scenario of this model and in which life was formed is that of underwater hydrothermal sources (*hydrothermal vents* known as *black fumaroles*) at the bottom of the oceans. In the model, all organic compounds were formed in situ. We are faced with self-sufficient metabolism, which is called *autotrophic*. In the primordial soup hypothesis, on the contrary, the first living beings were *heterotrophic*, namely, organisms that, to survive, must refer to organic compounds previously synthesized by other organisms, which are instead called *autotrophs*, such as all plants. The energy and "reducing power" necessary to transform carbon monoxide and dioxide into organic matter are due to the formation of pyrite, a compound of iron and sulphur, which starts from compounds of sulphur, iron and hydrogen. This thesis was confirmed experimentally in 1990. After this, Wächtershäuser proposed a whole series of reactions that started with the assimilation of carbon monoxide and carbon dioxide and ended with the generation of cells. Furthermore, current metabolic pathways would have been preceded by a series of reactions not accelerated by enzymes. The first point to explain is how inorganic carbon is incorporated. According to the author, this was possible owing to an autocatalytic carbon dioxide fixing process. This cycle promotes the fixation of carbon dioxide into organic molecules. Wächtershäuser called his hypothesis the *iron–sulfur world*. Stanley Miller argued that hydrothermal vents are too hot and that the heat destroys the chemicals produced, such as amino acids. Geologist Mike Russell reported fossil evidence of thermal vents with temperatures below 150 °C in the 1980s. Furthermore, the fossil remains of these vents contained pyrite, and he suggested that the first complex organic molecules formed within the pyrite structures. Russel suggested that thermal vents in the deep sea, which are warm enough to allow pyrite structures to form, harbor Wächtershäuser organisms. According to this thesis, life begins at the bottom of the sea, and metabolism appears first. Starting from the ideas of Peter Mitchell, Russell concluded that the ideal place for the formation of life is a hydrothermal vent with alkaline water. Therefore, the acidic *Corliss vents*, in addition to being too hot, would not have worked for this purpose. The first alkaline hydrothermal vents were discovered by Deborah Kelley in the Atlantic in a location called Lost city. The water temperature is between 40 and 75 °C and is slightly alkaline. These chimneys were perfect for Russell's ideas (who, in the meantime, began collaborating with the biologist William Martin), who became convinced that these were, in reality, the places where life was born. The rocks of the vents are porous and form pockets containing pyrite among various chemical substances. Given the natural proton gradient from the vent, these locations are ideal locations for metabolism to begin. After life harnesses the chemical energy of vent water, it begins producing molecules such as RNA. With the formation of the membrane, a true cell is formed, which then moves from the porous rock to the open sea. Proponents of the RNA world have found two problems with this theory. The first problem is that there is no experimental evidence for the processes described by Russell and Martin. The results are expected from experiments by Nick Lane, who hopes to observe metabolic cycles and perhaps even RNA. The second problem is the location of thermal vents in the deep sea and the fact that long-chain molecules such as RNA and proteins cannot form in water without enzymes.

In the last decade, a third approach has appeared promising as a way to create an entire cell from scratch.

Cells from the Beginning

In the RNA world, membranes are not present, although it is clear that it is unlikely that RNAs would be found in solution without protection. In the hypothesis of Wächtershäuser, the membranes, even if they appeared, did so late. Researchers have become aware that it is difficult to imagine life forms without having membranes. In current biology, metabolism, genetics and cellularity are closely linked. Cellularity depends on membranes, which are not simple semipermeable barriers but also have important metabolic capabilities and are fundamental for energy generation. Given the objective difficulties, the researchers chose to attempt to obtain the three aforementioned characteristics separately. Given the failure of the approach, a new trend arose to obtain them simultaneously, or at least to obtain two of the characteristics: cellularity and a second of the characteristics mentioned. Michael Russel underlines the importance of iron–sulfur membranes, which, according to him, form bubbles and microcavities in fumaroles in the deep ocean. We have also demonstrated the importance of membranes in the energetics of cells. In today's cells, membranes are made up of lipids and proteins. Lipids are the essential element for closing vesicles because they are molecules that have a polar and a nonpolar part. The molecules can associate with each other and self-organize through nonpolar zones, placing the polar zones in an aqueous medium that is nonpolar. The membranes grow, the vesicles swell, and at a certain point, they divide spontaneously. In addition to their ease of formation and ability to form microenvironments, vesicles can generate proton concentration differences between two sides of the membrane. As shown by Deamer in 2015, through cycles of water intake and loss, lipid vesicles give rise to the formation of chains of nucleotides. During the water intake phase, the RNAs are incorporated into the vesicles. Previous findings have led several researchers to propose a *world of lipids* that precedes RNA lipids.

Jack Szostack set the goal of achieving RNA replication together with the growth and reproduction of vesicles, i.e., an RNA cell capable of evolving. This last story is linked to a collaboration between Szostack and the champion of the idea of "first compartmentalization", namely, Pier Luigi Luisi. The ideas of the latter can be traced back to those of Oparin's coacervates. Luisi's challenge was to create protocells, but despite various experiments, he failed to create anything truly realistic and convincing. In 1994, he suggested that the first protocells must contain RNA, which must be able to replicate within the protocell. This idea quickly gained the support of Jack Szostak. The latter, in 2001, achieved great success. He added small amounts of some sort of clay into his experiments, and this sped up the rate of vesicle creation by a factor of 100. Furthermore, the latter absorbed the RNA strands from the surface of the clay. The following year, Szostak's team discovered that protocells could grow on their own. Was it possible that they could even reproduce? In 2009, Szostak and his student Ting Zhu created protocells with several concentric walls. By providing fatty acids, they grew and took on a thread-like shape. A gentle shear force allowed the protocell to shatter into dozens of daughter protocells that contained the parent protocell's RNA. There was one thing left to do: make the RNA replicate. Orgel spent much of the 1970s and 1980s studying how RNA strands were copied. To do this, one needs to use a single strand of RNA together with loose nucleotides. Nucleotides are used to assemble a second strand of RNA, which is complementary to the first. Orgel discovered that, under certain circumstances, RNA strands can be produced without any help from enzymes. This may have been related to how early life created copies of its genes. In a 2013 study, Orgel and his student, Kataryna Adamala, managed to realize Luisi's proposal and carry out RNA replication inside fatty acid vesicles. Szostak's team managed to construct protocells that retain their genes and simultaneously absorb molecules from the outside. Protocells can grow and divide, and RNA can replicate inside. These results and those of Sutherland, already mentioned, suggest a new unified approach to the origin of life, which is based on all three functions—genetics, metabolism, and cellularity—could be achieved simultaneously.

Appendix C: Details on Habitable Planets

The planet discovered thus far with the highest ESI (0.95) is Teegarden b, which is located in the *conservative habitable zone*, i.e., the part of the habitable zone where favorable conditions remain such for a good part of life. Another planet with a lower ESI also rotates around the star: Teegarden c. Teegarden b was discovered in 2019 after 3 years of research with the radial velocity method, owing to the CARMENES spectrograph capable of finding small variations in radial velocity even in small stars. Without a high-precision instrument, it would have been difficult to detect it because of the star's position and low brightness. Teegarden b is more internal than its companion and has an orbital period of 4.91 days, a composition probably similar to that of Earth, and the presence of water. Teegarden b receives approximately 21% more radiation than Earth receives from the Sun. Scientists from the team who discovered the two planets believed that the average temperature was close to 28 °C. The greatest problem for the habitability of this planet is that its star is a red dwarf, and Teegarden b rotates very close to it in *synchronous rotation*, that is, always turning the same hemisphere toward the parent star. Furthermore, red dwarfs are subject to violent flares, especially in the youthful phase, but since the star is 8 billion years old, it seems to be quite stable. Teegarden c has an ESI of 0.68, much lower than that of the companion planet, and is more similar to Mars; however, having a greater mass, it is probably able to retain a dense atmosphere capable of increasing the temperature due to the greenhouse effect, as happens for Earth, which, despite an equilibrium temperature of −18 °C, has an average surface temperature of approximately 16 °C. In any case, it receives only 35% of the radiation that the Earth receives from the Sun; therefore, assuming an atmosphere similar to that of the Earth, it has a surface temperature estimated at −47° centigrade. It completes one orbit around the star in 11.4 days. A positive factor for the habitability of the planet is the fact that the star Teegerden, a red dwarf, as already mentioned, is quite stable.

TOI 700 d with an ESI of 0.93 was discovered by the TESS space telescope in 2020 via the transit method together with two other planets (TOI 700b and TOI 700 c) closer to the red dwarf TOI 700 in the conservative habitable zone, which is 101.5 light years away from us. The mass is estimated to be between 1.4 and 2 times greater than that of the Earth; therefore, it is a terrestrial planet. It is located at a distance of 0.163 astronomical units from the star and rotates around it in 37 days. The planet receives 86% of the radiation that Earth receives from the Sun, and there should be conditions to support liquid water on the surface, which is essential for the presence of life as we know it. TOI 700 d has the same problem as Teegarden b: being close to the star, it has a synchronous rotation, i.e., it always shows the same face as the star. Obviously, one hemisphere is always illuminated, and the other hemisphere is in darkness. Its temperature estimate without an atmosphere is −4 °C, and depending on the atmospheric composition, its temperature could be in the range of −33–77 °C, i.e., an average of 22 °C.

Kepler-1649 c is a rocky planet, similar to Earth in size, with a radius of 1.06 times that of Earth. It is located in the conservative habitable zone. It was discovered with observations between 2010 and 2013 by the Kepler observatory, and the discovery was announced in 2017. The planet orbits the red dwarf Kepler-1649 in 19.5 days. In addition to Kepler-1649 c, the planet Kepler-1649 b, similar to Venus, rotates around the star. Kepler-1649 c receives 75% of what the Earth receives from the Sun from the star. The surface temperature is likely similar to that of the Earth, but it is not known whether liquid water is present on the planet, as the composition of the atmosphere is not known. Since its star is a red dwarf, the typical flares of these red stars could severely hinder the development of life on the planet.

Around the red dwarf star 2MASS J23062928-0502285 40 light years from Earth, also known as Trappist-1, 7 planets rotate in the *conservative habitable zone*. The star is a red dwarf of unknown age. In some works, an age of approximately 500 million years has been estimated; in others, a completely different age has been reported. In the first case, this would be bad news for the star's planets due to the exuberance of the red dwarfs. Kepler and Spitzer space telescopes have revealed possible faculae (bright spots on the surface), which correlate with flares from stellar eruptions. Flares emit a large amount of energy, hundreds of billions of times greater than the most powerful nuclear bombs. Since the star's planets orbit at distances 10–100 times smaller than the Earth–Sun distance, they could be strongly affected, and life may not form on them. *TRAPPISTs-1 b* and *c* were observed by James Webb and have nonexistent atmospheres or atmospheres that are so rarefied that they are almost nonexistent.

TRAPPIST-1 d has an ESI of 0.91. It was discovered in 2016 along with two other planets, and in 2017, the other 4 planets of TRAPPIST-1 were discovered. It is less massive and somewhat smaller than Earth and is assumed to be rocky in nature. Assuming that it reflects light such as the Earth and neglecting the possible greenhouse effect, a temperature of approximately 17 °C was estimated. Studies from 2018 estimated a mass lower than that at the time of discovery, approximately 30% of that of the Earth, with a radius of 77%; therefore, a density lower than that of the Earth could indicate the presence of large quantities of water in the liquid state in the form of oceans. The same study suggested that the planet has a relative amount of water 250 times greater than that of Earth. To have clearer ideas about this planet, we must wait until it is observed by the James Webb telescope. *TRAPPIST-1 e*, a rocky planet announced in 2017, was discovered via the transit method. It is the fourth of the seven planets that orbit the star. It is exoplanetically similar in size to Earth. It has a radius of 0.92 Earth radii and a mass equal to 0.69 Earth masses. The planet takes just 6.1 days to complete an orbit around it and is likely (given its low eccentricity) also in synchronous orbit, meaning that it always has the same face as the star. This characteristic, according to some research, reduces, if not completely compromises, the habitability of the planet. However, the presence of a sufficiently dense atmosphere would allow the transport of excess heat from the illuminated face to the dark one, allowing the presence of liquid water on the surface, particularly in the areas along the *terminators* or *circles of illumination*, which is the fictitious line that delimits the illuminated part from the shadowed part. Recent studies have shown that in terms of size, composition and flux of radiation it receives from the star, it appears to be the most Earth-like Trappist-1 planet. Research from 2020, however, suggests the opposite of what has been said regarding habitability. According to the study, this would be the planet with the greatest probability of life in the TRAPPIST-1 system. This study estimates that 93% of the planet's habitable surface and that it is the most likely to have vegetation that would increase the habitable area of the planet even further, even reaching 100% of the total surface area. In fact, the average global temperature would be between 14 and 25 °C, the maximum temperature on the day side would be 47 °C, and the minimum temperature on the night side would be 2 °C. The planet, along with the entire planetary system of TRAPPIST-1, will be observed by the James Webb Space Telescope, which can resolve the different views on the planet. *TRAPPIST-1 f* has an ESI of 0.68. The dimensions are similar to those of the Earth: the radius of the Earth is 1.05, the mass is 0.93 times greater than the terrestrial mass, the density is slightly lower than the terrestrial density, and the rotation period is 9 days. A 2020 study indicated that the surface temperature is −70 °C, which implies the absence of liquid water on the surface. The planet could be equipped with an underground ocean, and it is possible that cryo-volcanism phenomena, i.e., the emission of cold material, could be present, whereas tidal forces could heat the interior, as in Enceladus and Europa. *TRAPPIST-1 g* has an ESI of 0.58, and it is a rocky planet. It is the largest of TRAPPIST-1's seven planets, has a rotation period of 12 days and is located approximately 7 million kilometers from the star. It receives only 26% of the radiation that the Earth receives from the Sun, and its temperature should be approximately −70 °C. Because it is more massive than Earth, it is possible that it maintained a dense atmosphere, which would have allowed it to create a greenhouse effect sufficient to heat the surface to the melting point of water.

LP 890--9c has an ESI of 0.89 and was discovered via the transit method in 2022. The planet rotates in the conservative habitable zone of the ultracold red dwarf LP 890-9. In the LP 809-9 system, there is another planet, *LP 890-9 b*. *LP 890-9 c* is a super-Earth; it is slightly larger than Earth and has a radius of 1.37 Earth radii, but the mass is not known precisely and should be less than 25.3 Earth masses. It orbits over a period of 8.45 days, at a distance of only 6 million kilometers. The temperature without considering the greenhouse effect is approximately −1 °C, which is higher than the Earth's temperature, which is −18 °C and reaches 15 °C, taking into account the greenhouse effect. Given the small distance from the star, the planet is certainly in synchronous rotation and therefore turns the same hemisphere toward the star, creating problems with habitability. The red dwarf LP 809-9 is 7 billion years old and is therefore expected to be quite stable without the flares typical of red dwarfs. The planet will probably be observed by the James Webb Space Telescope to study its atmosphere.

Proxima Centauri b (or more simply Proxima b) has an ESI of 0.87, orbits at 0.05 astronomical units (approximately one eighth of the distance that separates Mercury from the Sun) from the red dwarf Proxima Centauri, which is 4.22 light years away from Earth. It is located in the conservative habitable zone. It has a mass between 1.17 and 3 Earth masses. The discovery was announced in 2016. It is not known whether it is a rocky planet, and its composition and atmospheric conditions remain unknown since no transitions have been observed. Its mass suggests that it may be a terrestrial planet if its radius is around terrestrial values, whereas if it is 1.4 terrestrial radii, it is likely that it is completely covered by a single ocean 200 km deep. The planet receives 65% of the total light flux that the Earth receives from the Sun, but in the infrared region, only 2% of the radiation that the Earth receives from the Sun in the visible region and approximately 400 times the X-ray flux that the Earth receives from the Sun. It is not known whether the planet always has the same face as the star and therefore shows a notable difference in temperature between the two hemispheres or whether it is in conditions similar to those of Mercury, with an alternation of day and night and therefore a much less extreme environment and average temperatures more similar to those on Earth and, in this case, has liquid water on the surface. There is no certainty about habitability, but the fact that the parent star is a red dwarf implies two usual problems: synchronous rotation and strong flares. To have a habitable portion, a thick atmosphere is necessary to guarantee heat exchange between the two night and day areas (in the case of synchronous rotation). A 2017 study and subsequent studies on super flares observed on Proxima b led us to think that the planet is not the best candidate for searching for extraterrestrial life forms.

K2-72 e, is one of four planets discovered orbiting in the conservative habitable zone around the red dwarf star K2-72, 228 light years away from us. The discovery of the planet was confirmed in 2016 via data from the Kepler space telescope. The planet takes just over 24 days to complete one revolution, and the proximity to the star also means that K2--72 is likely rotating synchronously. This causes a temperature difference of tens of degrees depending on which side of the planet you are on. The ESI is 0.87, and it is possible that there is liquid water along the terminator zones.

Gliese 1002 b, (ESI 0.86) rotates in the conservative habitable zone around a red dwarf. Although the temperature is not very different from that of Earth, it presents the typical problems of planets around red dwarfs: synchronous rotation and star flares.

Three planets rotate around the star Gliese 1061, a red dwarf. Of the three planets in the system, *Gliese 1061 d*, which is in the conservative habitable zone, is the one most likely to have liquid water on the surface.

Ross 128 b (ESI 0.86) is a rocky exoplanet and was discovered in 2017, rotating around a not very active red dwarf and particularly close to Earth. Preliminary studies seem to suggest that the planet may be temperate (21 °C).

Gliese 273 b (ESI 0.85) revolves around a red dwarf in the conservative habitable zone. If it had an Earth-like atmosphere, the planet would have an average surface temperature of approximately 19 °C, which is very similar to Earth's average temperature. If it does not undergo synchronous rotation, the distribution of heat across the entire planetary surface would be more efficient, while the stability of its star may have allowed its atmosphere to persist for billions of years, unlike other planets orbiting dwarf red stars, which are often subjected to violent flares capable of wiping out the atmosphere and making a planet uninhabitable.

The ones I talked about are just some of the habitable planets, but they all revolve around red dwarfs, which, if not old enough, are problematic stars; furthermore, the planets are usually in synchronous rotation, and the light always reaches one hemisphere. Are there also planets around stars such as the Sun, i.e., yellow dwarfs (class G) or rather orange dwarfs (class K)? An example is *Kepler-452b, which was* discovered in 2015 owing to the Kepler space telescope. The ESI was set at 0.83. The planet is the first with dimensions similar to those of the Earth and orbits in the habitable zone of a star very similar to those of the Sun. It has a revolution period of 385 days; it was formed before our planet and has a mass of 5 Earth masses. NASA calls the planet "Earth's older cousin." If it were a rocky planet, it would be a super-Earth, and considering its mass, it would be geologically active with erupting volcanoes and covered, if seen from space, with a thick blanket of clouds. The star, presumably radiating approximately 10% more energy than the Sun due to its evolution, could have triggered a growing uncontrolled greenhouse effect similar to that which can be detected in the solar system on Venus. However, because the planet is 60% larger than Earth, it is likely that it could retain oceans for a longer period, preventing Kepler-452b from escaping the greenhouse effect for another 500 million years. Owing to the effects of volcanic activity, any potential surface life could inhabit the planet for another 500–900 million years before the habitable zone extends beyond the orbit of Kepler-452 b. Researchers from the SETI Institute are using a radio telescope in California to search for radio transmissions coming from Kepler-452 b.

Kepler-1638 b (ESI 0.76) orbits Kepler-1638, a star similar to the Sun in mass, temperature, age, and metallicity. The planet has an orbital period of 259 days, a radius of 1.87 times that of Earth, and is likely classifiable as a super-Earth. It should be within the habitable zone.

Kepler-442 b (ESI 0.84) is an Earth-type planet that orbits around the conservative habitable zone of the orange dwarf Kepler-442, a K-type star. Orange dwarfs remain stable for much longer than yellow dwarfs do, which is why they are often cited as the best candidates around which habitable planets could exist. The planet has a radius of 1.34 times that of the Earth and a mass between 2.34 and 2.64 Earth masses. It revolves around its parent star in 112.31 days, at an average distance of 0.409 astronomical units. At that distance, the planet is likely not rotating synchronously. The probability that the composition of the exoplanet is made up mainly of rock and iron, such as Venus and Earth, is high, greater than 60%. The planet is within the habitable zone of its star, even closer to the center of the habitable zone, and its HZD (Habitable Zone Distance) is therefore better than that of our planet, whose orbit is further shifted toward the boundary interior of the habitable zone of the Sun. For the planets discovered with the transit method, another index was introduced, the HITE (Habitable Index for Transiting Planets), which gives great importance to the eccentricity of the orbit and the albedo, i.e., the ability of a surface to reflect light. Kepler-442 b has a HITE of 0.836 greater than that of Earth (0.829). Liquid water should be present on the surface. Its average temperature should be between 0 and −50 °C. However, being more massive, its atmosphere is likely to be denser, and consequently, the greenhouse effect is also greater than that of Earth, contributing to the rise of planetary temperatures up to an average temperature of 33 °C. Even with an atmosphere similar to that of Earth, large areas of the surface have temperatures above 10 °C. Furthermore, according to a 2015 study, Kepler-442 b has a degree of habitability even higher than Earth's (0.836 versus 0.829 on Earth), and depending on the atmospheric conditions (not yet known), it is the most serious candidate to be considered a *superhabitable planet*, i.e., a planet whose conditions for the development of life are more favorable than those on Earth.

Kepler-62 e (ESI 0.83) orbits the orange dwarf star Kepler-62. The planet, with a radius of 1.6 times that of the Earth, is probably a super-Earth with a solid surface and is located in the star's habitable zone, where the presence of liquid water on the surface is possible. It orbits its star once every 122 days at a distance of 0.427 astronomical units, along with the other 4 confirmed planets in its star system. Considering an atmosphere similar to that of Earth, the average temperature should be +29 °C.

Other planets revolving around orange dwarfs include *Kepler 1544b* and *Kepler 283 c*.

There are also planets in multiple systems. For example, *Gliese 667* is a multiple star system made up of two K-class stars, a little cooler than the Sun, and a red dwarf. In the system, there are several extrasolar planets, *Gliese 667 Cf* (ESI 0.76), which rotates in the conservative habitable zone around the red dwarf and has a temperature 34 degrees lower than the Earth's equilibrium temperature. Beyond this, there would also be the planet *Gliese 667 C* (ESI 0.60).

Kepler-296 e (ESI 0.85) is one of the 5 planets that rotates in the conservative habitable zone around the binary star Kepler 296, which is made up of an orange dwarf and a red dwarf. It is a terrestrial-type exoplanet. Discovered in 2014 as part of the Kepler mission, it is the fourth of the five planets discovered in the system. It is the smallest of the five planets discovered; however, it was initially from data from the Kepler telescope. It has a radius between 1.28 and 1.82 Earth radii, so it could be a gaseous dwarf planet without a solid surface.

Kepler-16 (AB) b is not of much interest because it is a gaseous planet without a surface; however, it has become famous because it is a circumbinary planet, namely, it orbits around the binary system Kepler 16. For this reason, referring to the Star Wars saga, which is shown a double sunset on the *planet Tatooine*, the planet was renamed Tatooine.

K2-18b revolves around a red dwarf and is approximately eight times the mass of Earth with a 33-day orbit. In 2023, K2--18 b was observed with the James Webb Space Telescope, which revealed the presence of carbon-containing molecules in its atmosphere, including methane and carbon dioxide and perhaps the molecule dimethyl sulfide. The abundance of these two molecules supports the hypothesis that the planet may be an ocean planet with a hydrogen-rich atmosphere. On Earth, the dimethyl sulfide molecule is produced only by life, especially phytoplankton. However, in May 2024, a series of simulations revealed that the signal relative to dimethyl sulfide is highly superposed to that of methan and that distinguishing between methan and dimethyl sulfide exceeds the ability of the James Webb Space Telescope. In any case, the atmosphere contains methane and carbon dioxide, which together are strong indicators that favor life.